Marta Hrabalova

Viscoelastic and thermal properties of natural fibre-filled composites

AF063491

Marta Hrabalova

Viscoelastic and thermal properties of natural fibre-filled composites

Potential of physical and chemical modifications for biodegradable composites

Südwestdeutscher Verlag für Hochschulschriften

Impressum/Imprint (nur für Deutschland/only for Germany)
Bibliografische Information der Deutschen Nationalbibliothek: Die Deutsche Nationalbibliothek verzeichnet diese Publikation in der Deutschen Nationalbibliografie; detaillierte bibliografische Daten sind im Internet über http://dnb.d-nb.de abrufbar.
Alle in diesem Buch genannten Marken und Produktnamen unterliegen warenzeichen-, marken- oder patentrechtlichem Schutz bzw. sind Warenzeichen oder eingetragene Warenzeichen der jeweiligen Inhaber. Die Wiedergabe von Marken, Produktnamen, Gebrauchsnamen, Handelsnamen, Warenbezeichnungen u.s.w. in diesem Werk berechtigt auch ohne besondere Kennzeichnung nicht zu der Annahme, dass solche Namen im Sinne der Warenzeichen- und Markenschutzgesetzgebung als frei zu betrachten wären und daher von jedermann benutzt werden dürften.

Verlag: Südwestdeutscher Verlag für Hochschulschriften GmbH & Co. KG
Dudweiler Landstr. 99, 66123 Saarbrücken, Deutschland
Telefon +49 681 37 20 271-1, Telefax +49 681 37 20 271-0
Email: info@svh-verlag.de

Approved by: Wien, BOKU, Diss., 2011

Herstellung in Deutschland:
Schaltungsdienst Lange o.H.G., Berlin
Books on Demand GmbH, Norderstedt
Reha GmbH, Saarbrücken
Amazon Distribution GmbH, Leipzig
ISBN: 978-3-8381-1080-6

Imprint (only for USA, GB)
Bibliographic information published by the Deutsche Nationalbibliothek: The Deutsche Nationalbibliothek lists this publication in the Deutsche Nationalbibliografie; detailed bibliographic data are available in the Internet at http://dnb.d-nb.de.
Any brand names and product names mentioned in this book are subject to trademark, brand or patent protection and are trademarks or registered trademarks of their respective holders. The use of brand names, product names, common names, trade names, product descriptions etc. even without a particular marking in this works is in no way to be construed to mean that such names may be regarded as unrestricted in respect of trademark and brand protection legislation and could thus be used by anyone.

Publisher: Südwestdeutscher Verlag für Hochschulschriften GmbH & Co. KG
Dudweiler Landstr. 99, 66123 Saarbrücken, Germany
Phone +49 681 37 20 271-1, Fax +49 681 37 20 271-0
Email: info@svh-verlag.de

Printed in the U.S.A.
Printed in the U.K. by (see last page)
ISBN: 978-3-8381-1080-6

Copyright © 2011 by the author and Südwestdeutscher Verlag für Hochschulschriften GmbH & Co. KG and licensors
All rights reserved. Saarbrücken 2011

Universität für Bodenkultur Wien

Department for Agrobiotechnology, IFA-Tulln

Institute for Natural Materials Technology
Head of Institute: Ass.Prof.Dr. Norbert Mundigler

Advisors:
Univ.Prof.Dr. Rupert Wimmer
Univ.Prof.Dr. Alfred Teischinger
Ass.Prof.Dr. Norbert Mundigler
Ao.Univ.Prof. Dr. Antje Potthast
Univ.Ass. Dr. Adriana Gregorova

VISCOELASTIC AND THERMAL PROPERTIES OF NATURAL FIBRE-REINFORCED COMPOSITES

Dissertation
for obtaining a doctorate degree at the University of Natural Resources and Life Sciences Vienna

Submitted by
Marta Hrabalova

Vienna, March 2011

Acknowledgements

It is an honour for me to express gratitude and thank to my advisor, Prof. Rupert Wimmer, for his leading role and great support during the entire time of my studies, and for securing important contacts and cooperations that contributed notably to this thesis. I am also grateful to my project associate and co-advisor, Dr. Adriana Gregorova, for the countless discussions and great professional as well as personal support she always provided.

I would like to thank the Institute for Natural Materials and Technology at IFA Tulln, particularly my co-advisor Prof. Norbert Mundigler, for providing such a good and friendly working environment, lab space and also technical support. Likewise, I thank the entire staff at the Institute of Wood Science and Technology, Department of Material Sciences and Process Engineering, and particularly my co-adviser Prof. Alfred Teischinger, for continued assistance in the many measurement and analyses activities. I owe my deep gratitude to Dr Manfred Schwanninger from the Institute of Chemistry for his continued assistance, discussions and support of my work. I am also grateful to Prof. Antje Potthast for co-advising my dissertation by being available with natural polymer expertise.

I am also grateful to Prof. Dr. Klaus Richter, former head of Wood laboratory at Empa Duebendorf, as well as to Dr. Tanja Zimmermann for allowing me to accomplish my internship at Empa. Thanks are due to Thi Thu Thao Ho for assisting me in the laboratory.

Thanks to my friends, who have been supporting me during difficult times. I will not forget. I will be always ready to give it back. As well I will never forget all strong good and clever people I have met. Thank you just for being and giving me the strength to believe that the right way is still not empty.

The last but the most I want to thank my family, especially my parents who faced out the hard times with me. They gave me the absolute freedom in my decisions and maximal possible support.

Abstract

This study is focused on viscoelastic and thermal properties of composites using thermoplastic biodegradable matrices, in combination with cellulose-based fibreous fillers. First, interactions among these components were studied by looking at responses of viscoelastic and thermal properties relative to processing conditions and physical or chemical modification of fillers and the polymer matrix. Effects of composite processing on fibre-matrix interactions were studied at different crystallization stages. Processing alterations were applied to softwood flour filled poly(lactid acid) films. Responses were investigated using differential scanning calorimetry (DSC), dynamic mechanical analysis (DMA), and scanning electron microscopy. It was found that thermal and viscoelastic properties can be improved with suitable thermal annealing. The second part was concerned with poly(lactic acid) filled with different tissue types (mature, juvenile, compression wood) of Sitka spruce. Wood fibers were treated with vinyltrimethoxysilane, while the polymer matrix was modified with 4,4-methylene diphenyl diisocyanate. It is shown that modification treatments improved thermal and mechanical properties. It was also demonstrated that there is potential to improve biobased composites by utilizing the natural variability of wood fibres.

Nano-fibrilated cellulose was produced from flax and wheat straw pulps through high pressure disintegration. The reinforcing potential of the nano-fibrils in a polyvinyl-alcohol matrix was evaluated. It is shown that the selection of the appropriate raw cellulose is indispensable, being essential for the functional optimization of composite products. Compared to the others the improvement with nano-fibrillated cellulose was the most remarkable one. This was assigned mainly to the structural features of this filler. It was also shown that different modification treatments were efficient with regards to viscoelastic and thermal properties. Finally, the influence of surface modification of beech wood flour was investigated in Poly (3-hydroxybutyrate)/wood flour composites. In addition to hydrothermal (HT) pre-treatment, sodium hydroxide and stearic acid were both used as surface modifiers. Effects on the adhesion of PHB/wood fibre interface were feeble when no hydrothermal pretreatment was applied. The surface modifiers applied on hydrothermally pretreated wood fibres significantly improved interface adhesion.

Keywords: biopolymer, composites, viscoelastic properties, thermal analysis, nano-fibrillated cellulose, wood fibre

Zusammenfassung

Diese Arbeit beschäftigt sich mit viskoelastischen und thermischen Eigenschaften von Verbundwerkstoffen, unter Verwendung biologisch abbaubarer, thermoplastischer Matrices, in Kombination mit cellulosischen Faser- Füllstoffen. Die Wechselwirkungen zwischen diesen Komponenten wurden als Erstes mittels viskoelastischer und thermischer Eigenschaften studiert, in Bezug auf Prozessbedingungen bzw. chemo-physikalischen Modifikationen von Füllstoff und Matrix. Auswirkungen der Prozessbedingungen auf die Faser-Matrix Anbindung wurden bei verschiedenen Kristallisationsgraden untersucht. Verschiedene Prozessbedingungen wurden bei Nadelholzmehl-gefüllten Polymilchsäurefilmen angewandt. Die Eigenschafts-änderungen konnten mittels differentieller Scanning-Kalorimetrie (DSC) bzw. Dynamisch-mechanische Analse (DMA) bzw. Elektronenmikroskopie ermittelt werden. Die thermischen und viskoelastischen Eigenschafte waren durch geeigneter Temper-Verfahren stark verbessert. Das zweite Kapitel beschäftigt sich mit Poly(Milchsäure), verstärkt mit drei unterschiedlichen Gewebetypen der Sitka-Fichte (juvenile und adules Holz, Druckholz). Die Holzfüllstoffe wurden weiter mit Vinyltrimethoxysilan bzw. die Polymer-Matrix mit 4,4-Diphenylmethan-Diisocyanat modifiziert. Es zeigten sich deutliche Verbesserungen der thermischen und mechanischen Eigenschaften. Die Verbesserung kann somit durch gezielte Auswahl des Holzrohstoffes bzw. durch die richtige Wahl des Modifikationsverfahrens erzielt werden. Der dritte Teil ist nano-fibrillierter Cellulose gewidmet, hergestellt aus Flachs und Weizenstroh-Zellstoffen durch Hochdruckdesintegration. Das Potential zur Verstärkung von Polyvinylalkohol-Matrices wurde untersucht. Es konnte gezeigt werden, dass die richtige Rohstoffwahl beim Hochdruckaufschluss ein wesentlicher Faktor ist, damit die Funktionalität des Verbundwerkstoffes verbessert werden kann. Die Verbesserung durch nanofibrillierte Zellulose war vergleichsweise am höchsten, was auf die speziellen Eigenschaften dieses Füllstoffes zurückzuführen ist. Die verschiedenen Modifikationen waren in puncto viskoelastischer und thermischer Eigenschaften sehr effektiv. Schließlich wurde der Einfluss der Oberflächenmodifikation von Buchenholzmehl in Poly (3-Hydroxybutyrat)/Holzmehl Kompositen untersucht. Zusätzlich zur hydrothermischen Vorbehandlung wurden auch Natriumhydroxid und Stearinsäure-Behandlungen angewandt. Die Faser-Matrix Anbindung war ohne hydrothermischer Vorbehandlung nur schlecht ausgebildet. Viskoelastische bzw. thermische Eigenschaften konnten durch diese Behandlungen verbesser werden.

<u>Stichworte</u>: Biopolymer, Verbundwerkstoff, viskoelastische Eigenschaften, thermische Analyse, Nano-fibrillierte Zellulose, Holzfaser

Table of content

1 **Introduction** _____ 9
   1.1 Background _____ 9
      1.1.1 Modifications _____ 10
         1.1.1.1 Matrix modification _____ 11
         1.1.1.2 Filler modifications and treatments _____ 11
      1.1.2 Characterisation of viscoelastic and thermal properties _____ 12
2 **Objective and definition of the research topic** _____ 16
3 **Materials and Methods** _____ 16
   3.1 Materials _____ 166
      3.1.1 Polymer matrixes _____ 16
      3.1.2 Reinforcements _____ 17
   3.2 Chemicals and modifications _____ 17
   3.3 Composite preparation _____ 17
   3.4 Characterisation _____ 20
      3.4.1 Dynamic mechanical analysis _____ 20
      3.4.2 Differential scanning calorimetry _____ 20
      3.4.3 Infrared spectroscopy _____ 20
      3.4.4 Scanning electron microscopy _____ 21
      3.4.5 Viscosimetric measurement _____ 21
      3.4.6 Mechanical testing _____ 21
      3.4.7 Water-uptake _____ 211
4 **Results and discussion** _____ 22
   4.1 Preparation method _____ 22
   4.2 Chemical modification of polymer matrix _____ 24
   4.3 Physical disintegration of the fillers _____ 25
   4.4 Chemical modification of the filler _____ 27
   4.5 Filler origin _____ 30
5 **Summary and Conclusions** _____ 32
6 **Table of abbreviations** _____ 35
7 **References** _____ 37
8 **Index of tables** _____ 41
9 **Table of figures** _____ 42
10 **Selected publications** _____ 43
11 **CV** _____ 86

Appendix

This thesis is based on four publications, which appear in peer-reviewed scientific journals.

A **Hrabalova, M.**, Gregorova, A., Wimmer, R., Sedlarik, V., Machovsky, M. & Mundigler, N. (2010); Effect of wood flour loading and thermal annealing on viscoelastic properties of poly(lactic acid) composite films; *Journal of Applied Polymer Science*, 118, 3, 1534–1540.

B Gregorova, A., **Hrabalova, M.**, Wimmer, R., Saake, B. & Altaner, C. (2009); Poly(lactide acid) Composites Reinforced with Fibers Obtained from Different Tissue Types of Picea sitchensis; *Journal of Applied Polymer Science*, 114, 2616–2623.

C **Hrabalova, M.**, Schwanninger, M., Wimmer, R., Gregorova, A., Zimmermann, T. & Mundigler, N., (2010); Fibrillation of Flax and Wheat Straw Cellulose and Its Effect on Thermal, Morphological and Viscoelastic Properties of Poly(vinylalcohol)/Fibre Composites; *Bioresources*, 6, 2, 1631-1647.

D Gregorova, A., Wimmer, R., **Hrabalova, M.**, Koller, M., Ters, T. & Mundigler, N. (2009); Effect of Surface Modification of Beech Wood Flour on Mechanical and Thermal Properties of Poly (3-hydroxybutyrate)/Wood Flour Composites; *Holzforschung*, 63, 565-570.

1 Introduction

1.1 Background

Biodegradable polymers are research and industry attracting materials due to intentions of solving waste disposal issues that arise from the sustainability of synthetic polymer waste. So far, synthetic biodegradable polymers are most commonly used in medical and pharmaceutical services, but might find new applications in other industrial branches as well, partially replacing petroleum-based polymers. Biodegradable polymers are divided into three groups: (1) biopolymers of natural origin, (2) synthetic biodegradable polymers such as poly(vinyl alcohol) (PVA), and (3) modified polymeric materials to reach biodegradability.[1] The more regular use of biodegradable synthetic polymers such as poly(lactic acid) (PLA), which is the first commodity polymer produced from renewable resources[2], or poly(hydroxy butyrate) (PHB), is limited due to drawbacks such as brittleness, low glass transition and a higher price.

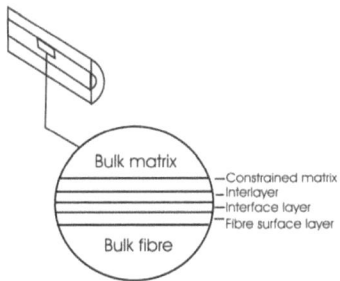

Figure 1: Interface between fiber and matrix (adapted from: Christopher J. Spragg et al. Fiber, Matrix and Interface Properties)[3]

Especially the low glass transition temperature (T_g) of PLA is intensely discussed since around 60°C segmental motions are released, which dramatically lowers mechanical properties, making PLA less applicable in industrial applications.[4] A similar dramatic change in properties with temperature was observed for PVA.[5] Since these changes take place close at application temperatures T_g practical applications of these materials are limited.

Temperature dependant segmental molecular motion releases affecting mechanical properties are gradual in the case of PHB, but mechanical performance is lowered as well.[6]

Particulate composite materials are composed of at least two solid phases: (1) a continuous polymer matrix, and (2) a discontinuous fibre reinforcement. Ideally, the composite materials outbalance the properties of their individual constituents. Synthetic biodegradable polymers are used as matrixes to obtain composite material with acceptable performance. Frequently used reinforcements for synthetic biodegradable polymers are cellulose-based fibres. Cellulose is renewable, naturally abundant, low-cost, and has good mechanical properties combined at low density. Consequently, it is expected that composites formed from biodegradable thermoplastic polymers and cellulose-based fibrous filler will have high specific stiffness and strength, with improved temperature performance over the neat matrix polymer.

The fibrous filler is supposed to act as a mechanical constraint to polymer matrix especially in interface region (Figure 1). Polymer matrix at the interface with the filler has higher stiffness than the bulk matrix because of reduced polymer chain mobility and different packing density of the macromolecules.[3] Interfacial bonding strength between reinforcement and polymer matrix is also based on molecular polarity of components. Moreover the polarity of components is also closely linked to the ability of the fibrous filler to be homogeneously distributed throughout the matrix within the processing. Composite properties vary with quantity, mechanical properties, shape as well as spatial distribution of the fibrous filler in the polymer matrix, combined with the interaction between the components (adhesion, bonding). Strength of interaction can be modified by chemical modification of the filler, or alternatively by modification of the polymer matrix.

1.1.1 Modifications

The good adhesion between the fibrous filler and the polymer matrix is one of the fundamental aspects for optimal performance of composites. Mostly, the adverse affinity of hydrophobic matrix and the hydrophilic filler causes poor interaction, low homogeneity and imperfect stress transfer from polymer matrix to the fibrous filler.[7] This problem covers also composites with cellulose fibrous filler.

The incompatibility of cellulose with the polymer matrix might be suppressed by filler or polymer matrix modification. Physical modification (i.e. thermal treatments, drawing, high pressure fibrillation, low temperature plasma, sputtering or corona discharge etc.) of polymer matrix or cellulose fibrous fillers results in the change of surface properties of the fiber and may improve interaction with polymer matrix. Chemical modifications are done primarily to introduce new functional groups (grafting hydrophobic or bifunctional molecules), to cellulose fiber or chain of polymer matrix to enhance the compatibility. Fillers become more hydrophobic, consequently an agglomeration in hydrophobic matrix is avoided by reaching higher homogeneity.[8] The most common chemical coupling methods for cellulose filled composites are silanization or use of isocyanates.[7]

1.1.1.1 Matrix modification

The thermal history of composites experienced during processing is an important factor affecting their later performance, as it influences the physical nature of the polymer matrix. Physical constraints are closely related to the polymer morphology which impacts markedly resulting macroscopic features such as mechanical and thermal properties of polymer consequently of composite.[9] One way of influencing the polymer crystalline morphology is thermal annealing. There are two known crystallization mechanisms: crystalline growth, and crystalline thickening. The crystalline growth is evoked at temperatures lower than crystallization temperature and the mechanism of the crystalline thickening is initialized through thermal annealing taking place above the crystallization temperature. Thermoplastic polymers often possess several crystal modifications leading to complex melting behaviour.[10-13] Besides, fillers may act as nucleating agents in the polymer matrix.[14] Consequently, the interfacial performance between the matrix and the fibrous filler is driven by character of transcrystalline region, promoted on the fibrous filler surface.[12, 14, 15] **Publication A** illustrates the changing interaction between PLA and wood flour (WF) in composites with altering annealing temperature which was applied within the melt processing of composites in temperature range between 90 to 120°C.

Another way of modifying the polymer matrix is through chain extenders. They are bifunctional substances and their addition increases the elastic modulus and the heat resistance glass transition of the polymer.[16] Diisocyanates are the most commonly applied chain extenders for aliphatic polyesters.[17] 4, 4-methylene diphenyl diisocyanate (MDI) with highly reactive hydroxyl groups is able to react with carboxyl and hydroxyl groups by forming urethane linkages,[18] which improves for example also the thermal stability of the material or its solubility. (**Publication B**)

1.1.1.2 Filler modifications and treatments

Coupling agents are used to enhance fibrous filler dispersion within the matrix and promote adhesion and bonding between fillers and matrix. Silane coupling agents such as organosilane compounds have at least two reactive groups. Various types of silanes are conventionally used as coupling agents to improve compatibility between hydrophilic cellulosic filler and hydrophobic polymer matrix and they also act as surfactants (agents used in small amounts in composites to improve filler dispersion).[14, 19] Vinyl trimethoxysilane

(VTMO) was used as surfactant to modify wood flour from Sitka spruce (Picea sitchensis), which was used as a filler in a PLA matrix. (**Publication B**)

Stearic acid is another potential surfactant for mineral fillers as well as wood flour composites.[20-23] **Publication D** is particularly concerned with stearic acid treatment of beech wood flour, in combination with hydrothermal modification to avoid fibre agglomeration in PHB matrix.

Mechanical disintegration of cellulose fibrils into smaller cellulose constituents via high pressure fibrillation provided material with nanometer dimension, high aspect ratio, increased specific area and excellent mechanical properties.[24] Compared to the non-fibrillated precursors, nano-fibrillated cellulose (NFC) is supposed to provide a higher number of OH-groups available for hydrogen bonding.[25, 26] NFCs produced from flax pulp and wheat straw pulp in mass fraction ranging between 5 to 40% was used as reinforcement in a poly(vinyl alcohol) (PVA) matrix. (**Publication C**)

It is well known that hydrothermal treatment causes structural, chemical, physical and mechanical changes of wood properties.[27] Due to hydrothermal treatment the amorphous phase in wood is degraded by increasing the rate of the crystalline phase.[28-30] Solid wood as well as WF from beach were both hydrothermally treated with natrium hydroxide and used to incorporate PHB polymer matrix. Alkaline treatment (treatment with natrium hydroxide solution) changes the chemical composition of wood as well as physical properties.[31] Alkaline treatment is accompanied by the removal of hemicelluloses, lignin and waxes. Through this the wood fibres are roughened and fibrillated and their surface area is enlarged. The modified fibres are expected to have better fiber-matrix adhesion when used in composites.[32] (**Publication D**)

1.1.2 Characterisation of viscoelastic and thermal properties

Thermoplastic polymeric material is described by viscous and elastic parameters.[33] Elastic component induces re-forming of the material into its initial shape. Material deformation upon applied stress is ascribed to viscous component, while elastic deformation is attributed to bond stretching within the polymer chain. Viscous flow is ascribed to chain slipping, i.e. disconnection of intermolecular interactions.[33] During deformation and stress release some energy input is stored and recovered, while other is dissipated as heat. This behaviour is called viscoelastic. The response of the sample to stress provides information on material stiffness and ability to dissipate energy.[34]

Dynamic mechanical analysis allows observation of viscoelastic properties of polymer materials responding to temperature and frequency changes. DMA measurement can be done in different modes such as tension, compression, shear or bending. Sample is subjected to sinusoidally varying load/harmonic oscillation. Energy dissipation is revealed as the phase angle δ between the applied stress and the measured strain (Figure 2).

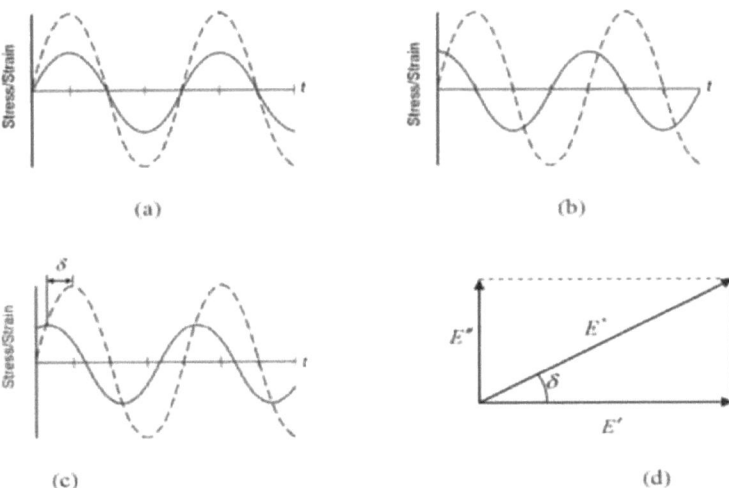

Figure 2: Phase angle between stress and strain amplitudes within DMA measurement for a) elastic, b) viscous, c) visco-elastic and d) diagram of complex modulus in complex plane.[35]

Perfectly elastic material displays no phase shift so the deformation occurs in phase with the applied stress (Figure 2a). Contrary to this the phase difference for perfectly viscous material will respond with the lag of δ = π/2 behind the applied stress (Figure 2b). Figure 2c demonstrates the phase difference for linear viscoelastic material. As seen in Argand diagram (Figure 2d) due to the phase shift δ between stress and strain its modulus is introduced as complex modulus

$E^* = E' + iE''$ (1)

and the phase shift can be determined as

$\tan \delta = \dfrac{E''}{E'}$. (2)

tan δ provides the information how well material gets rid of the energy and it is reported as "tangent of phase", "loss tangent" or "damping". The peak amplitude of measured stress corresponds to the storage modulus (E') which is the amount of recoverable energy stored in the sample. Loss modulus (E'') is the amount of irrecoverable energy dissipated as a heat.[19, 35]

Consequently, energy dissipation depends on the material and for composites it strongly depends on the composition of the sample in case of blends, and on the strength of interaction. Dissipation of energy is different in composite system than in its precursor. Also the quality of the stress transfer from the matrix to the filler might be detected with DMA. The weak interfacial bond will dissipate more energy than a strong interfacial bond. In other words, improved bonding might be shown by lowering of loss tangent.[3] Moreover, DMA is suitable tool for T_g detection and makes possible to observe the changes in material mechanical properties over the wide temperature range. Determination of T_g by DMA is preferred above other thermal analysis methods (particularly DSC) due to the fact that larger sample sizes can be measured at higher accuracy. Moreover, evaluation of T_g by DMA is easier due to the greater change in obtaining signals arising from T_g.[36] Signal changes can be seen with E' curve-onsets, with inflection points of downward E' slopes, with $tan\ \delta$ maxima, and with E'' peaks It is also evident that each method delivers slightly different T_g value,[37] and therefore one method has to be consistently used with a given set of samples (see Figure 3).

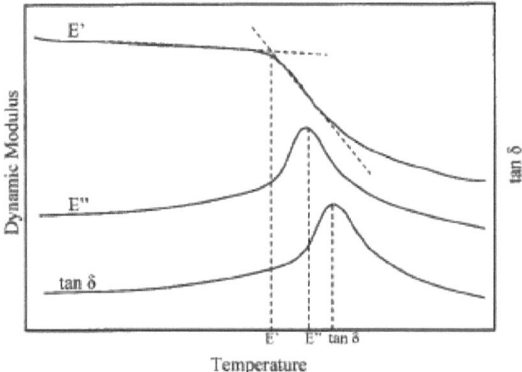

Figure 3: Example for T_g determination from DMA output demonstrating the production of different values obtained in dependency of method used. Onset of the E', peak of E'' and $tan\ \delta$. (Adapted from: Herzog et al. (2005) Glass-transition temperature based on dynamic mechanical thermal analysis techniques as an indicator of the adhesive performance of vinyl ester resin.)[37]

Thermal characteristics of thermoplastic materials are equally important as mechanical. As it is known, properties of thermoplastic polymer alter with temperature due to the polymer chain reorganization. The most important parameters of semicrystalline polymers are threshold temperatures at which the physical properties of polymer material abruptly change: melting temperature (T_m), crystallization temperature (T_c) and T_g.

Differential scanning calorimetry is a common thermoanalytical technique that measures the change of the difference in the heat flow rate between a sample and the reference sample, as a function of temperature or time under heating, cooling or isothermal conditions.[38] Generally, the temperature program for a DSC analysis is designed in a way that the sample temperature increases linearly with time. The reference sample should have a well-defined heat capacity over the measured temperature range.[39-41]

The basic principle underlying this technique is that a sample undergoing a physical transformation such as phase transitions, projected as exothermic or endothermic processes, heat is given or needed to maintain equal temperature for sample and reference.[39-41]

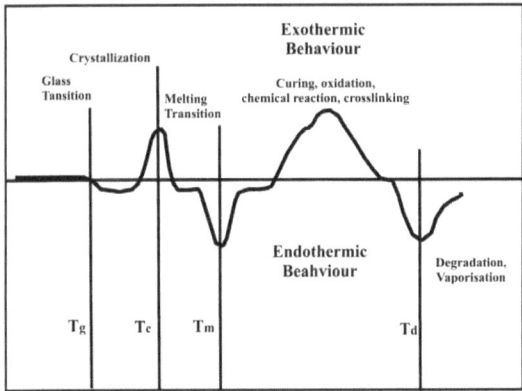

Figure 4: The typical DSC output for thermoplastic semi-crystalline polymer (adapted from: Dubief et al. (1999) Polysaccharide microcrystals reinforced amorphous poly(beta-hydroxyoctanoate) nanocomposite materials).[43]

With a DSC it is possible to characterise thermal properties of thermoplasts such as melting and crystallization temperatures, and the glass transition temperature. However, as mentioned DSC is not considered as the most reliable method for T_g determination. In DSC output T_g is displayed as the step change in the baseline, which indicates a change in heat capacity). This step change is often small and might be expressed as a gradual shift. Therefore, T_g value determined by DSC have limited reliability.[36]

Moreover, amount of energy absorbed/released enthalpy of fusion (ΔH), consequently crystallinity (χ_c), or degree of thermal degradation due to exothermic or endothermic responses can be indicated (Figure 4). In multi-component polymer systems interactions between matrix and the filler will have an effect on the thermal properties.[41]

2 Objective and definition of the research topic

The five hypotheses of this work are 1) Processing conditions will notably affect compatibility between fiber and the matrix in thermoplastic composite. 2) Different wood tissue types are sources of natural variability, and included as a cellulose-based filler in thermoplastic polymer matrix they will have a prominent effect on viscoelastic properties of the resultant composite. 3) Compatibility between fibrous filler composite might be improved by using the proper chemical modification of both, the fibrous filler as well as the polymer matrix. 4) The disintegration of fibrous fillers into their nano-scaled constituents will uniquely influence mechanical and thermal properties of polymer composites. 5) With the combination of hydrophilic polymer matrix and the hydrophilic fibrous filler the resulting material reaches unique properties due to intermolecular interaction between filler and matrix, in particular viscoelastic properties.

3 Materials and Methods

3.1 Materials

3.1.1 Polymer matrixes

PLA 7000D pellets with PLA density of 1.24 g/cm^3 were received from NatureWorks LLC (Minnetonka, MN). (**Publication A, B**) PVA Airvol 523 (medium viscosity, degree of hydrolysis 86 to 88% and average molecular weight ~ 89,000) was supplied by AIR Products and Chemicals, Inc. (USA). (**Publication C**) PHB was received as a homopolymer powder from Graz University of Technology, Institute of Biotechnology and Biochemical Engineering, Graz, Austria. (**Publication D**)

3.1.2 Reinforcements

Commercially available softwood flour (SWF) was supplied by J. Rettenmaier & Söhne GmbH & Co., Germany. SWF was ground and homogenized to finally pass a 120 μm sieve (**Publication A**). Sitka spruce [*Picea sitchensis* (Bong.) Carrie`re] Juvenile wood (JW), mature wood (MW), and compression wood (CW) were isolated from 36-year-old tree grown in Kershope (Northumbria, UK) (**Publication B**). For the isolation of NFC two commercially available cellulose pulps from wheat straw (WS) and flax (F) were used. (**Publication C**) Technical WF from European beech (*Fagus sylvatica* L.) with a particle size of 120 mm was supplied by Lindner Mobilier s.r.o. Madunice, Slovakia. (**Publication D**)

3.2 Chemicals and modifications

- Silane treatment was applied on each tissue type of Sitka spruce fiber. VTMO was obtained from Fluorochem, Derbyshire UK. Modification was accomplished in 0.5 wt % VTMO solutions in methanol. (**Publication B**)
- PLA chain extension was provided by 4, 4-methylene diphenyl diisocyanate (obtained from Sigma Aldrich, Germany). The modification with MDI was accomplished according to Li.[18] (**Publication B**)
- Stearic acid was obtained from Carl Roth, Germany. Stearic acid (SA) treatment was applied on beech wood and hydrothermally treated beech wood with 0.07 M stearic acid solution in toluene (Carl Roth, Germany). (**Publication D**)
- Alkali treatment of beech wood flour and hydrothermally treated beech wood flour was undertaken in 5% w/v aqueous solution of sodium hydroxide (Carl Roth, Germany). (**Publication D**)
- High pressure fibrillation was accomplished to produce NFC from flax and wheat straw cellulose pulps. Microfluidizer, M-110Y was utilized for high pressure fibrillation which came after proper cellulose swelling, grinding and inline dispersion by small ultra-turrax. (**Publication C**)
- Hydrothermal pre-treatment of beech wood took place under hot and steamy conditions in a laboratory - scale autoclave. (**Publication D**)

3.3 Composite preparation

Composite films were prepared according to Figure 5. The term "modified polymer matrix" stands for crystallized PLA matrix or PLA chains extended with MDI; the "modified reinforcement" in this work stands for the reinforcement after chemical as well as mechanical modification.

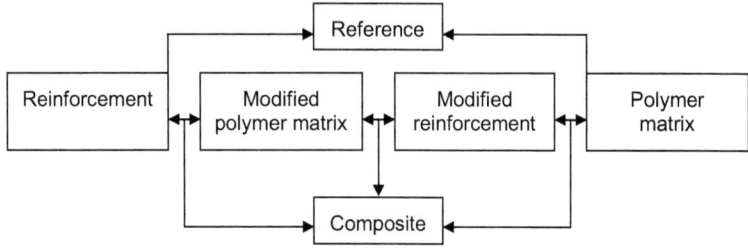

Figure 5: Scheme of the work

Polymer blends based on PLA with different proportion of SWF (0-50% SWF mass fraction in PLA) were prepared by melt blending in a Brabender kneader at 190°C and 20 rpm. As soon the PLA melted in the kneader the wood flour was added, and the mixture was blended for 5 minutes. The composites were moulded to thin films in a hot press. For thermal annealing the moulded films were kept in press at annealing temperatures of 90, 100, 110, and 120ºC, respectively, for 45 min. (**Publication A**)

Neat PLA, PLA/MDI and PLA/Sitka composites were prepared via the solution casting method. For this 1.2g of PLA was solved in 50mL of chloroform before the appropriate amount of Sitka fibers was added. Mass fraction of the wood flour in each mixture was 20%. Blends were casted on Petri dishes and dried (1 day air dried, followed by 2 days vacuum drying). Dry films were disintegrated and melt moulded in the hot press. (**Publication B**)

Equally PVA/NFC composites were prepared via solution casting method. Aqueous PVA solution was mixed with appropriate amount of flax or wheat straw NFC in aqueous suspension such that the NFC mass fractions in PVA matrix ranged between 0 to 40%. Mixtures were casted on a Teflon plate and dried. (**Publication C**)

Films from PHB and PHB/beech wood flour composites were prepared by repeated compression moulding. (**Publication D**)

Table 1 represents the blend compositions, modifications and appropriate labelling.

Table 1: Sample labelling.

Sample	Matrix	Matrix modification	Filler	Filler modification
PLA_0_90	PLA	Thermal annealing 90 °C	-	-
PLA_0_100	PLA	Thermal annealing 100 °C	-	-
PLA_0_110	PLA	Thermal annealing 110 °C	-	-
PLA_0_120	PLA	Thermal annealing 120 °C	-	-
PLA_20_0	PLA	-	20% SWF	-
PLA_20_90	PLA	Thermal annealing 90 °C	20% SWF	-
PLA_20_100	PLA	Thermal annealing 100 °C	20% SWF	-
PLA_20_110	PLA	Thermal annealing 110 °C	20% SWF	-
PLA_20_120	PLA	Thermal annealing 120 °C	20% SWF	-
PLA_10_0	PLA	-	10% SWF	-
PLA_30_0	PLA	-	30% SWF	-
PLA_50_0	PLA	-	50% SWF	-
PLA/JW	PLA	-	20% JW Sitka spruce	-
PLA/MW	PLA	-	20% MW Sitka spruce	-
PLA/CW	PLA	-	20% CW Sitka spruce	-
PLA/JW/silane	PLA	-	20% JW Sitka spruce	Silane treatment
PLA/MW/silane	PLA	-	20% MW Sitka spruce	Silane treatment
PLA/CW/silane	PLA	-	20% CW Sitka spruce	Silane treatment
PLA/JW/MDI	PLA	MDI	20% JW Sitka spruce	-
PLA/MW/MDI	PLA	MDI	20% MW Sitka spruce	-
PLA/CW/MDI	PLA	MDI	20% CW Sitka spruce	-
PVA_20F-NFC	PVA	-	20% F-NFC	Physically refined
PVA/20WS-NFC	PVA	-	20% WS-NFC	Physically refined
PVA/40F-NFC	PVA	-	40% F-NFC	Physically refined
PVA/40WS-NFC	PVA	-	40% WS-NFC	Physically refined
PHB/WF	PHB	-	20% WF beech	-
PHB/WF/AT	PHB	-	20% WF beech	Alkaline
PHB/WF/SA	PHB	-	20% WF beech	Stearic acid
PHB/HT-WF	PHB	-	20% WF beech	Hydrothermal
PHB/HT-WF/AT	PHB	-	20% WF beech	Hydrothermal + alkaline
PHB/HT-WF/SA	PHB	-	20% WF beech	Hydrothermal + stearic acid

3.4 Characterisation

3.4.1 Dynamic mechanical analysis

Viscoelastic properties of composites were measured on a NETZSCH 242 C in tensile mode. Strips of 10 mm × 6 mm × 0.1 mm in size were cut. tan δ, E'', E' values were measured at a strain sweep frequency of 1 Hz, and a heating rate of 3 °C/min. The temperature range is individual for each material. T_g was evaluated either from the E' inflection point or peak temperature of tan δ. (**Publications A, B, C, D**)

3.4.2 Differential scanning calorimetry

Melting behaviour of composites was studied with a NETZSCH DSC 200 F3 Maja. Up to 10 mg of sample in plate form were placed into an aluminium pan covered by a pierced lid. Measurements were done at a heating rate of 10 K/min (**Publications A, B, C**) or alternatively at 20K/min (**Publication D**), under continuous nitrogen gas flow (20 or 60 mL/min). T_m and enthalpy of melting (ΔH_m) were derived from the endothermic melting peak and T_c with crystallization enthalpy (ΔH_c) was evaluated from the exothermic peak. One heating scan was sufficient to characterise the status of the sample used for DMA. The degree of polymer matrix crystallinity is proportional to the ΔH_m and it is calculated according to the following equation:

$$\chi_c (\%) = \frac{\Delta H_m}{\Delta H_m^0} \times \frac{100}{w} \qquad (3)$$

where, w is the actual mass fraction of PLA matrix, ΔH_m the experimentally determined heat of fusion, and ΔH_m^0 the heat of fusion of a 100% crystalline sample (theoretical value). ΔH_m° = 146 J/g was used for the PHB homopolymer,[44] and ΔH_m° = 93.7 J/g was used for PLA.[45] (**Publication A, B, C, D**)

3.4.3 Infrared spectroscopy

Attenuated total reflection-Fourier transform infrared spectra (ATR-FTIR) of non-fibrillated and nano-fibrillated celluloses as well as the spectra of PVA/NCF films were measured with 32 scans per sample at a spectral resolution of 4 cm^{-1} and a wavenumber range between 4000 cm^{-1} to 600 cm^{-1}. The attenuated total reflectance device was from MIRacle™, (Pike Technologies, www.piketech.com), with a DLATGS mid-infrared detector, installed on a Bruker Vertex 70 (Bruker Optics, www.brukeroptics.de). (**Publication C**)

Beech wood flour and chemically modified beech wood flour were embedded in KBr pellets and analyzed with a Vertex 70 FT-IR spectrometer (Bruker Optik) equipped with a Miracle-

Diamond (Pike). Spectra were collected in the range between 4000 and 600 cm^{-1} with an accumulation of 32 scans and a resolution of 2 cm^{-1}. (**Publication D**)

3.4.4 Scanning electron microscopy (SEM)

PLA/WF composite cryo-fractured surfaces were coated with thin layer of Au/Pd. Thermionic-emission scanning electron microscope (TESCAN VEGA/LMU) operated under high-vacuum mode at an acceleration voltage of 5 kV. (**Publication A**)

Aqueous wheat straw and flax suspensions of nano-fibrillated and non-fibrillated celluloses and cryo-fractured composite films were sputtered with a 7.5 nm thick platinum layer (BAL-TEC MED 020 modular high vacuum coating systems, BALTEC AG, Principality of Liechtenstein). Microscopic analysis was accomplished using a Jeol 6300F (Jeol Ltd., Japan), operated under high vacuum at 5kV beam voltage. Fractured surfaces of the PVA/NFC composites were studied with a scanning electron microscope, FEI Nova NanoSEM 230 also operating under 5kV. (**Publication C**)

The instrument used was the Tesla BS 300 SEM. All samples were coated with gold prior to the examination of morphology. (**Publication D**)

3.4.5 Viscosimetric measurement

The intrinsic viscosity $[\eta]$ of both non-fibrillated and fibrillated flax (F-NFC) and wheat straw cellulose (WS-NFC) was determined according to ISO 5351 standard (Anonymus, 2004). Degree of polymerization (DP) was then calculated using Staudinger-Mark-Houwink equation.

$$[\eta] = (K \times DP)^a \qquad (4)$$

where the values of the constants K and a depend on the polymer-solvent system (Henriksson, 2008). Values for WS were $K = 2.28$ and $a = 0.76$, while for F $K = 0.42$ and $a = 1$ (Marx-Figini, 1978). (**Publication C**)

3.4.6 Mechanical testing

Tensile strength, elongation at break, and Young's modulus of the samples were determined on a 100 N Zwick, Type BZ1 mechanical testing machine. A 25 mm grip clearance was used. Crosshead speed was 2 mm/min. (**Publication B, D**)

3.4.7 Water-uptake

Relative water uptake (*wt.%*) measurements were performed with the nano-fibrillated cellulose composites, the PVA and the PVA/NFC composite films, with 20 mm x 20 mm, in size, and a thickness of about 0.20 mm. Dry weight of each sample was about 100 mg. After dry weighing the samples were placed in a room conditioned at 50% relative humidity and 23°C. Weight changes samples were repeatedly measured up to 9 days, and water-uptake was calculated using the formula,

$$wt.\% = 100 * \frac{W_2 - W_1}{W_1} \qquad (5)$$

where, W_1 is the weight of the dry sample, and W_2 the weight of the wet sample.

4 Results and discussion

4.1 Preparation method

Thermal properties of polymer matrix (ΔH_m as well as T_m) have greatly changed with the annealing temperature as well as the WF content. Double melting behaviour was detected solely by samples annealed at 90 and 100°C.

It was evident that the capability of the material to store energy has increased as soon WF was incorporated (Figure 6). T_g also raised when WF was present (e.g. for PLA_0_0 T_g=63°C and for PLA_50_0 T_g was determined to be 68°C). Thermal annealing did not greatly affect the storage modulus in the glassy region, however, distinct changes were detected in rubbery region where E' gained higher values with WF incorporation. The reinforcing effect of WF for the storage modulus in the rubbery region was greatly accelerated with thermal annealing. A transcrystalline region has formed with WF, which is an important factor influencing adhesion between WF and PLA. Friction between polymer matrix and WF changed with the applied annealing temperature. With the DSC results and $\tan \delta$ it was possible to confirm the improved interaction between polymer matrix and the filler at annealing temperatures of 90 and 100°C. This result suggests that transcrystalline regions formed by crystalline growth mechanism improved interfacial bonding between PLA and WF.

In Figure 7 the filler effectiveness is shown as the constant C in dependency to WF content and the annealing temperature. The lower C is, the higher the effectiveness of the filler in polymer matrix. Results confirmed the influence of annealing temperature on the crystallization mechanism, and on the adhesion between WF and PLA matrix. At higher annealing temperatures crystalline thickening will dominate, (120°C), which ultimately lowers the filler effectiveness (Figure 7).

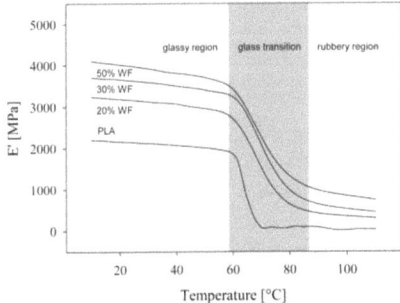

Figure 6: An example of DMA results: PLA with 0 to 50% of softwood flour annealed at 90°C improvement of E' in glassy and rubbery region due to the WF increasing content.

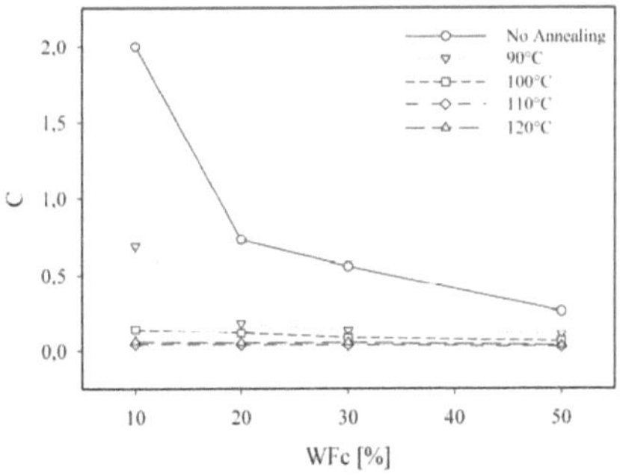

Figure 7: Effectiveness of the filler versus WF content and annealing temperature.

Table 2: Percentage changes of storage modulus in rubbery region (80°C) caused by thermal annealing related to the unannealed PVA or PVA/WF reference of relevant WF concentration.

Annealing temperature	WF mass fraction [%]				
	0 %	10 %	20 %	30 %	50 %
90°C	200	1000	2000	2000	1000
100°C	200	4000	2000	2000	1000
110°C	0	4000	2000	2000	1000
120°C	0	4500	2000	1000	1000

An example of the effect of WF filling thermally annealed films at 90°C on storage modulus is seen in Figure 6. E' was improved with WF mass fraction in each region. Table 2 gives the values of E' shift in the rubbery region of annealed samples relative the unannealed reference. Improvement due to the thermal annealing was most distinct at low WF concentrations. 4000 times higher E' was detected at annealing by 100°C with 10% WF mass fraction than for unannealed PLA/WF film. Considering the rubbery region the highest E' in rubbery region was reached with 50% WF filling (**Publication A**), but the rate of change caused by thermal annealing was relatively small. Interestingly, the improvement of viscoelastic properties were not found to be linked to the PLA crystalline fraction since χ_c was similar across the whole WF concentration and annealing range.

Table 3: Viscoelastic, thermal and tensile properties of PLA, PLA/WF and PLA/MDI composite films with 20% mass fraction of juvenile, mature and compression Sitka spruce WF.

Samples	DMA				Tensile tests		DSC	
	E' [MPa] 20 °C	E' [MPa] 80 °C	tan δ	T_g [°C]	Tensile strength	Young's modulus [GPa]	T_m [°C]	X_c [%]
PLA	2.88	0.15	0.474	48	45.8±4.7	2.62±0.17	150	18
PLA/JW	3.02	0.40	0.249	57	35.2±4.0	2.93±0.40	151	29
PLA/MW	3.74	0.51	0.210	61	45.3±7.3	3.39±0.24	150	25
PLA/CW	3.21	0.26	0.297	56	42.0±5.2	2.92±0.26	150	28
PLA_MDI/JW	3.45	0.12	0.526	59	39.6±2.8	2.78±0.22	150	6
PLA_MDI/MW	4.72	0.66	0.621	66	55.0±3.9	3.50±0.23	152	14
PLA_MDI/CW	3.85	0.18	0.503	60	46.2±4.2	3.10±0.23	150	6

The interesting side aspect of **Publication B** demonstrates well the importance of the used preparation method. As known the glass transition of PLA is reported at around 60°C in dependency on the PLA type.[2] However, by using the same PLA type as in **Publication A** T_g of neat PLA in **Publication B** turned out to be markedly lower T_g=48°C (Table 3). This result was confirmed by the T_g obtained from the tan δ peak, where higher values were expected than the one obtained from the E' inflection point. This discrepancy can be attributed to the residual chloroform in PLA matrix, which was acting as a plasticizer. This evidence is further supported by the E' values measured in the rubbery region. Using solution casting the E' at 80°C of neat PLA was much lower compared to melt-moulded PLA (0.15 MPa versus 250MPa). Thus, it is expected that by melt moulding prepared composites would have much better mechanical as well as thermal properties than composites prepared by solution casting.

4.2 Chemical modification of polymer matrix

MDI chain-extended PLA was reinforced with JW, MW and CW, with a mass fraction of WF in PLA_MDI of constantly 20%. Compared to composites with unmodified PLA matrix E' was improved especially in the glassy region. Tan δ increased due to the formation of branched disordered molecular structures. Melting temperature remained unchanged. The results of DMA, DSC and tensile tests are shown in Table 3. The effect of MDI modification on composite properties is remarkable. Viscoelastic properties of PLA_MDI modified composites are dependant on the WF tissue type (juvenile, compression, mature wood), The best improvement of PLA_MDI was detected for composites filled with mature wood WF. This indicates significant improvement of interfacial compatibility by the applied coupling agent. It has to be noted, that the effect of the MDI treatment on viscoelastic properties in the glassy region is more distinct than the effect of WF alone. The rubbery region was improved only by PLA_MDI/MW. The low thermal stability of PLA_MDI/wood flour composites might be assigned to lower crystallinity of polymer matrix. The lowering of crystallinity was indicated by DSC measurement. Melting temperature remained unchanged. (**Publication B**)

4.3 Physical disintegration of the fillers

Physical modification of fibres was accomplished via high-shear homogenization for flax and wheat straw cellulose pulps to obtain nano-sized cellulose fibril aggregates. The process of disintegration is described in **Publication C**. It was demonstrated that production and properties of the final NFCs are strongly dependent on the degree of polymerization of the raw material. SEM micrographs of raw materials as well as their nano-fibrillated analogues are shown in Figure 8. NFCs were used as reinforcements up to 40% mass fraction for hydrophilic PVA matrix. F-NFC was better dispersed in the PVA matrix than WS-NFC, which might be related to the finer structure of flax cellulose.[45] Results of DMA revealed better values of E' in the glassy region for wheat straw composites. This might be due to higher DP of WS-NFC. Mechanical properties in rubbery region were higher for PVA/F-NFC composites. In both cases T_g increased in dependency of the NFC content. Results of E' are depicted in Figure 9.

The changes of melting temperatures as well as heat of fusion of PVA in dependency of the NFC concentration are shown in Figure 10. The effect of the crystalline fraction on T_m and E' is eliminated. Thermal properties of PVA matrix in composites have changed after NFC insertion so that T_m decreased compared to the neat PVA reference, which is a consequence of interactions between the components in the system.[46, 47] It was concluded that the improvement of viscoelastic properties is a function of the filler fineness and its dispersion. The interaction between the OH groups of PVA and OH groups of the cellulose is an important factor as well. These interactions were confirmed by ATR-FTIR. It has to be noted that the samples were stored in a desiccator prior to measurements to avoid the plasticizing of PVA.

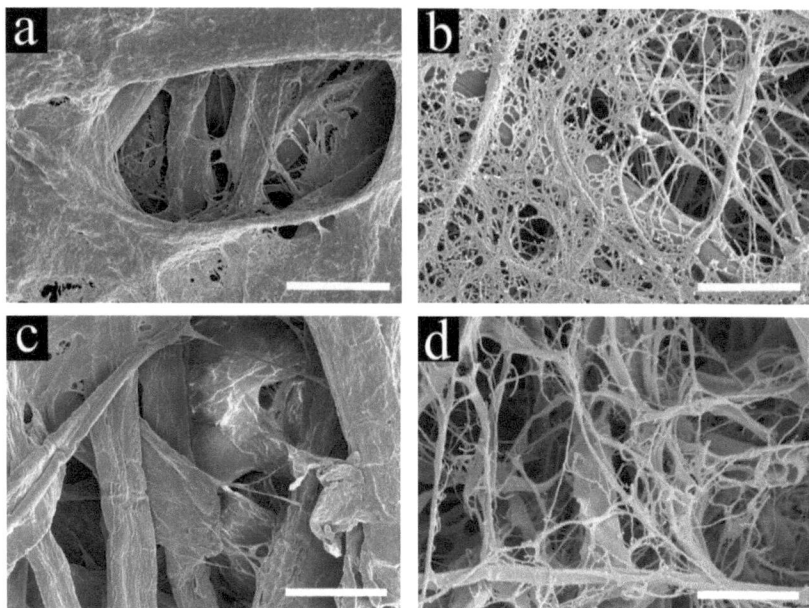

Figure 8: Scanning electron micrographs of a) nonfibrillated flax, b) flax NFC, c) nonfibrillated wheat straw and d) wheat straw NFC. White scale bars represent the length of 20 μm (a and c) and 3 μm (b and d).

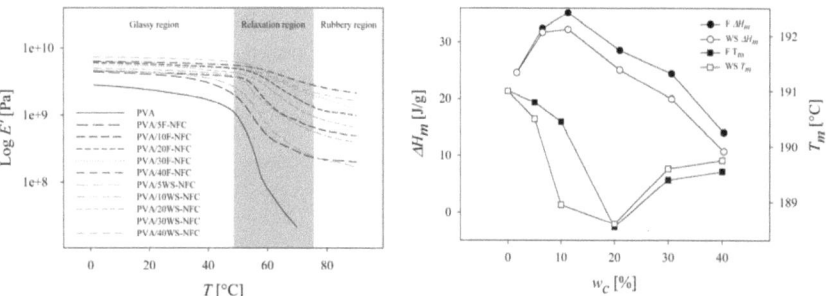

Figure 9: Improvements of PVA viscoelastic properties due to the NFC presence (1 Hz).

Figure 10: DSC results: melting temperatures and heat of fusion for PVA/F-NFC and PVA/WS-NFC over the w_c=5-40% NFC mass fraction range.

Since we have studied water susceptible system the sensitivity of materials at mild humid conditions was tested. The results revealed that PVA uptakes more moisture at given conditions than the cellulose films. Figure 11 shows results and values for water uptake can be estimated according to the "rule of mixtures",

$$P_H = P_1 V_1 + P_2 V_2 \qquad (6)$$

where, P_H is the water uptake of the composite, P_1 and P_2 are here water uptake of PVA and NFCs, respectively; and V_1 and V_2 are PVA and NFC volume fractions.[48] Figure 11 well illustrates the measured and the predicted water uptake. NFC cellulose increases water sensitivity of the PVA film, especially at low NFC fractions. The discrepancy between predicted and measured values can be assigned to different water-uptake mechanisms in the composite systems which are evoked by different intermolecular interactions.

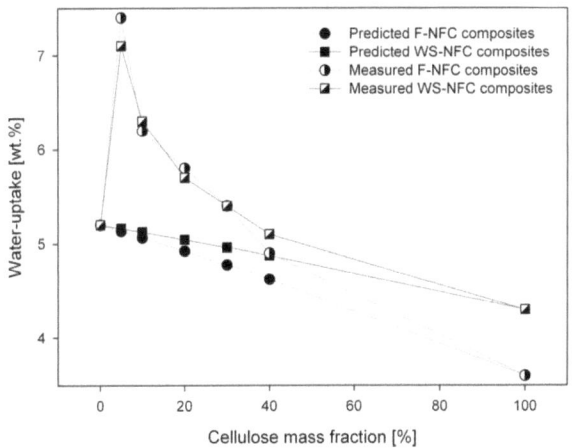

Figure 11: Water-uptake of cellulose, PVA and PVA-NFC composites, measured and predicted using the rule of mixtures.

4.4 Chemical modification of the filler

Juvenile wood, mature wood and compression wood flours from Sitka spruce were treated with silane and used as reinforcements in PLA matrices. Viscoelastic properties (E' at 20 and 80 ℃) did improve due to silanization up to 10% in the glassy region. Rubbery region remained more or less unchanged. Tan δ decreased especially for the silanized CW composites, which were showing improved adhesion between WF and the polymer matrix. T_g shifted to higher temperatures due to silanization up to 6℃, compared to the unmodified references. Crystallinity decreased at the same time. Both shifts were most likely caused by the crosslinking between PLA and silane treated WF.[49] Silane treated MW and CW has also improved the tensile strength and Young's modulus of the composite films. Results of DMA, DSC and tensile testing are shown in Table 4. (**Publication B**)

Table 4: Sitka spruce reinforced PLA matrix: Viscoelastic, thermal and tensile properties (all tensile test results were derived from 5 experimental runs).

Method	DMA				Tensile tests		DSC	
Samples	E' [MPa] 20 ℃	E' [MPa] 80 ℃	tan δ	T_g [℃]	Tensile strength	Young's modulus [GPa]	T_m [℃]	X_c [%]
PLA/JW	3.02	0.40	0.249	57	35.2±4.0	2.93±0.40	151	29
PLA/MW	3.74	0.51	0.210	61	45.3±7.3	3.39±0.24	150	25
PLA/CW	3.21	0.26	0.297	56	42.0±5.2	2.92±0.26	150	28
PLA/JW_silane	3.33	0.35	0.251	63	38.0±3.1	3.12±0.41	149	23
PLA/MW_silane	3.93	0.62	0.201	62	51.5±6.6	3.94±0.30	149	20
PLA/CW_silane	3.50	0.43	0.205	62	47.8±7.5	3.16±0.44	150	25

Beech wood flour, hydrothermally treated and alkaline treated, as well as stearic acid beech treated wood flour, was tested together with PHB matrices. Addition of modified WFs improved tensile properties of the composites. T_g of composites with modified WF was slightly shifted to higher temperatures. Composites filled with stearic acid and hydrothermally treated wood flours demonstrated 10 and 16% better E' in the glassy and rubbery region, respectively, which went hand in hand with enhanced tensile properties. Moreover, the T_m of composites with hydrothermally treated beech WF decreased most probably due to the interaction between modified filler and polymer matrix (**Publication D**).

As seen in Table 5 the major impact on PHB/WF composite properties was achieved with the hydrothermal treatment of WF, in combination with stearic acid treatment. Chemical treatments did not markedly contribute to the improvement of composite viscoelastic properties. This is also demonstrated by Figure 12. Stearic acid treatment improved the tensile properties of both composite types.

Table 5: Beech wood flour reinforced PHB matrix – results from DMA, DSC and mechanical testing.

Method	DMA				Tensile tests		DSC	
Samples	E' [MPa] 20 °C	E' [MPa] 80 °C	tan δ	T_g [°C]	Tensile strength	Young's modulus [GPa]	T_m [°C]	X_c [%]
PHB/WF	3.67	1.84	0.051	18	30.9±2.9	3.16±0.25	186	53
PHB/AT-WF	3.65	1.69	0.058	21	32.2±1.1	3.24±0.09	185	53
PHB/SA-WF	3.73	1.84	0.058	21	39.6±1.7	3.42±0.08	189	53
PHB/HT-WF	4.26	2.07	0.047	19	33.8±2.3	3.32±0.18	180	50
PHB/HT-WF-AT	4.26	2.13	0.053	18	32.8±2.9	3.31±0.03	181	54
PHB/HT-WF-SA	4.68	2.41	0.051	20	36.5±2.9	3.45±0.26	178	49

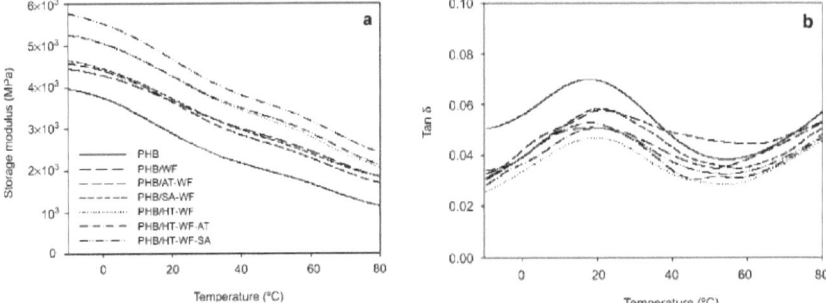

Figure 12: DMA results of PHB with 20% of modified and unmodified beech wood flour. a) increase in storage modulus due to the beech WF reinforcement and WF modifications, b) decrease and shift of loss tangent due to the reinforcement.

Table 6 summarizes the effects of chemical treatments of WFs on composite properties. The results are given as percentage changes in storage modulus in E' (at 20 and 80°C), tan δ, tensile properties and crystallinity. The changes are related to PLA and PHB composites having unmodified WF. The most pronounced improvements were reached with the silane treatment. Storage modulus was improved in the glassy as well as in the rubbery region, with the T_g shifting to higher temperatures. At least 10% positive shifts were detected also for tensile properties of the composites. Use of silane treated WF for PLA was accompanied by decreasing T_m and crystallinity.

Table 6: Relative changes in crystallinity, viscoelastic and tensile properties, and changes of T_g and T_m for PLA and PHB composites due to the chemical treatment of the filler. Changes relate to composites having unmodified fillers. (* T_m and T_g changes are given in ℃)

Method	DMA				Tensile tests		DSC	
Samples	E' [Δ%] 20 ℃	E' [Δ%] 80 ℃	tan δ [Δ%]	T_g^* [Δ℃]	Tensile strength [Δ%]	Young's modulus [Δ%]	T_m^* [Δ℃]	X_c [Δ%]
PLA/JW_silane	10	-10	0	6	10	10	-1	-20
PLA/MW_silane	5	20	0	1	10	20	-1	-20
PLA/CW_silane	9	70	-30	6	10	10	0	-10
PHB/AT-WF	0	-10	10	3	0	0	8	0
PHB/SA-WF	0	0	10	3	30	10	-1	0
PHB/HT-WF-AT	0	0	10	-1	0	0	0	10
PHB/HT-WF-SA	10	10	10	1	10	0	-3	0

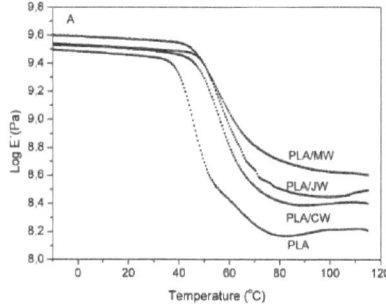 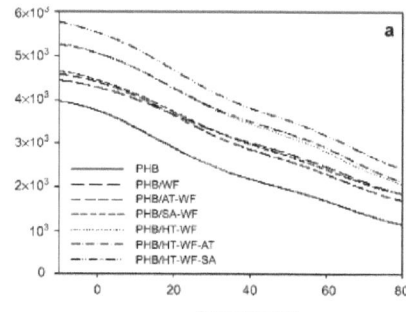

Figure 13: A) storage moduli of PLA/Sitka composites. Amount of Sitka spruce juvenile wood, mature wood and compression wood flour was 20%. Apparent storage modulus improvement in the rubbery region; B) tan δ of PLA/Sitka composites. Decrease of loss tangent and shift to higher temperature is dependent on the type of WF used.

4.5 Filler origin

The most pronounced change in viscoleastic properties within the WF filled composites seems to be with PLA plus Sitka spruce WF (**Publication B**). The use of different tissue type resulted in alteration of composite properties which is well demonstrated by DMA (Figure 13). The best improvement was detected using mature wood as a filler, with E' in rubbery region improved by 250%, and T_g by about 13℃ compared to neat PLA. On the contrary, poor results were reached with SWF reinforced PLA (sample PLA_20_0) reported in **Publication A**. Overall, the best viscoelastic property improvement was reached with the use of both types of NFCs in PVA matrix (**Publication C**) However, these results are not comparable with other results without considering the fact that the nature of the PVA matrix is polar in contrast to PLA and PHB.

5 Summary and Conclusions

In this work the viscoelastic and thermal properties of fibrous filler/thermoplastic biodegradable polymer based composites were studied as related to polymer morphology as well as to various chemical and physical treatments.

The effect of processing conditions was demonstrated by composite systems of different wood flour mass fractions in a PLA matrix. Thermal annealing of hot pressed moulded films resulted in a great improvement of composite viscoelastic properties in rubbery region. Compatibility between wood fibrous filler and PLA matrix was driven by transcrystralline regions, which was also highly influenced by the annealing temperature, which was linked to crystallization mechanisms. The glass transition region was broadened with higher amounts of fillers and also due to thermal annealing. Loss of mechanical integrity in this region was not as dramatic as for neat-unannealed PLA. This behaviour occurred due to physical changes in the polymer matrix. Fibrous fillers were also modified by thermal treatment.

Two different cellulose pulps derived from flax and wheat straw were homogenized under high pressure, achieving nano-sized cellulose fibrils. These fibrils were subsequently incorporated in a poly(vinyl alcohol) matrix. The pulp types differed primarily in degree of polymerization-which was not dramatically altered by the applied disintegration treatment. The PVA composite filled with the wheat straw nanofibrils had a higher degree of polymerization, and it delivered an improvement in the glassy region. On the other hand nano-fibrillated flax cellulose performed better in rubbery region. The finer the material, the easier it is to disperse in the PVA matrix. This fact was seen as the reason for the better performance in the rubbery region.

Marked differences were detected in composites filled with different tissue type fibres from Sitka spruce, in combination with a conventional silane treatment, and also a filler – matrix treatment using 4, 4-methylene diphenyl diisocyanate. Mechanical strength differences 'were related to the applied modification as well as to the used tissue types (mature wood, compression wood, juvenile wood). Mature wood was found to have the best potential as filler in PLA composites, in combination with the applied chemical treatment.

Chemical treatments were also applied to beech wood flour, and to hydrothermally treated beech wood flour. These materials were incorporated PHB matrices. Here, it was shown that the effectivity of alkali and stearic acid treatment was amplified by a hydrothermal pre-treatment.

Table 7: Results of DMA, DSC and tensile measurements: Storage moduli (E'), elongation at break (ε), tensile strength (TS), Young's modulus (E).

Filler	DMA				Tensile test			DSC	
	E' at 20°C*	E' at 80°C*	tan δ*	T_g•	TS*	ε*	E*	T_m•	ΔH_m*
PLA_0_90	-20	220	10	0	20	0	20	0	960
PLA_0_100	0	160	-70	2	10	0	20	3	1500
PLA_0_110	-10	-20	-40	0	30	10	30	-2	2290
PLA_0_120	-10	20	-20	0	10	-30	40	2	2280
PLA_20_0	0	40	-40	1	0	-30	10	2	3420
PLA_20_90	20	**2450**	-90	3	-10	-50	30	5	6110
PLA_20_100	10	**2370**	-150	5	0	-50	40	2	5700
PLA_20_110	10	**2310**	-90	2	-10	-60	40	3	5460
PLA_20_120	10	**2730**	-90	2	-10	-50	40	5	5670
PLA_10_0	0	0	-10	2	10	-30	10	1	450
PLA_30_0	20	110	-50	1	0	-50	30	2	3990
PLA_50_0	50	470	-70	5	10	-60	30	1	460
PLA/JW	10	170	-50	9	-20	-50	10	1	30
PLA/MW	30	250	-60	13	0	-60	30	0	10
PLA/CW	10	80	-40	8	-10	-40	10	0	20
PLA/JW/silane	20	140	-50	15	-20	-60	20	-1	0
PLA/MW/silane	40	320	-60	14	10	-30	50	-1	-10
PLA/CW/silane	20	190	-60	14	0	-40	20	0	10
PLA/JW/MDI	20	-20	10	11	-10	-40	60	0	-70
PLA/MW/MDI	60	350	30	18	20	-50	30	2	-40
PLA/CW/MDI	40	20	10	12	0	-40	20	0	-70
PVA_20F-NFC	**140**	**8000**	-60	8	-30	-60	20	-5	10
PVA/20WS-NFC	**130**	**5800**	-40	7	-40	-60	20	-4	-50
PVA/40F-NFC	**160**	**18200**	-80	10	-30	-70	50	-6	0
PVA/40WS-NFC	**200**	**14000**	-60	7	-10	-50	60	-4	-60
PHB/WF	30	60	-30	0	-20	-40	10	5	-20
PHB/WF/AT	30	50	-20	3	-10	-30	10	4	-20
PHB/WF/SA	30	60	-20	3	10	-20	10	8	-20
PHB/HT-WF	50	80	-30	1	-10	-40	10	-1	-30
PHB/HT-WF/AT	50	90	-20	0	-10	-40	10	0	-20
PHB/HT-WF/SA	60	110	-30	2	0	-30	20	-3	-30

* The results are given as percentage change in composite properties related to the neat material.

• Change in temperatures (T_g and T_m) are given in °C.

In summary, change rates of composite properties are given relative to neat and unmodified references (Table 8). To achieve improvements of properties a wide range of materials and treatments were used. It was found that the applied factors have considerable effects on several composite properties, and they need to be considered in further research efforts. However, there are still a number of possible chemical and physical treatments, which were not included in this work.

By proper choice of materials and treatments as well as combinations of both it is possible to achieve improved properties of materials that will gain more importance in a future bio-based economy It is expected that ongoing research in polymer synthesis will go hand to hand with proper processing technology, and with suitable additives and fillers the obtained biodegradable materials might reach competitiveness at acceptable costs.

6 Table of abbreviations

ATR-FTIR	Attenuated total reflection-Fourier transform infrared spectroscopy
C	effectiveness of fillers on the storage moduli
CW	compression wood
DMA	dynamic mechanical analysis
DP	degree of polymerization
DSC	differential scanning calorimetry
E'	loss modulus
E''	loss tangent
E^*	complex modulus
F	flax
F-NFC	nano-fibrillated flax cellulose
ΔH	enthalpy of fusion
ΔH_m	the enthalpy of melting
$\Delta H_m°$	the enthalpy of melting of 100% crystalline sample
ΔH_c	enthalpy of crystallization
HT	hydrothermal treatment
JW	juvenile wood
MDI	4, 4-methylene diphenyl diisocyanate
MW	mature wood
NFC	nano-fibrillated cellulose
PHB	poly(hydroxyl butyral)
PLA	poly(lactic acid)
PVA	poly(vinyl alcohol)
SA	stearic acid
SEM	scanning electron microscopy
SWF	softwood flour
T_g	glass transition
T_m	melting temperature
T_c	crystallization temperature
tan δ	loss tangent
VTMO	vinyltrimethoxysilane
WF	wood flour
WS	wheat straw
WS-NFC	nano-fibriallted wheat straw cellulose
$[\eta]$	intrinsic viscosity
X_c	crystallinity

7 References

1) Kaplan, D. L. (1998). *Biopolymers from Renewable Resources*. Berlin: Springer; Academic press.

2) Mohanty, A.K., Misra, M. & Drzal, L.T. (2005). *Natural Fibers, Biopolymers, and Biocomposites*. Broken Sound Parkway NW, Suite 300, Boca Raton, Florida, USA: Taylor & Francis Group.

3) Spragg, C.J. & Drzal, L.T. (1996). *Fiber, Matrix and Interface Properties*. Scranton, PA: ASTM International.

4) Yang, S., Wu, Z.-H., Yang, W. & Yang, M.-B. (2008). Thermal and mechanical properties of chemical crosslinked polylactide (PLA). *Polym Test*, 27, 957–963.

5) Roohani, M. Habibi, Y. Belgacem,N.M., Ebrahim, G., Karimi, A.N. & Dufresne, A. (2008). Cellulose whiskers reinforced polyvinyl alcohol copolymers nanocomposites. Eur Polym J, 44, 2489–2498.

6) Singh, S., Mohanty, A.K., Sugie, T., Takai, Y. & Hamada, H. (2008). Renewable resource based biocomposites from natural fiber and polyhydroxybutyrate-co-valerate (PHBV) bioplastic. *Compos Part A*, 39, 875–886.

7) George, J., Sreekala, M. S., & Thomas, S. (2001). A review on interface modification and characterization of natural fiber reinforced plastic composites. *Polym Eng Sci*, 41, 9, 1471-1485.

8) Huda, M.S., Drzal, L.T., Mohanty, A.K. & Misra, M. (2006). Chopped glass and recycled newspaper as reinforcement fibers in injection molded poly(lactic acid) (PLA) composites: A comparative study. *Compos Sci Technol*, 66, 1813-1824.

9) Garlotta, D., (2001). A Literature Review of Poly(Lactic Acid). *J Polym Environ*, 9, 63-84.

10) Zhang, J., Duan, Y., Sato, H., Noda, I., Yan, S. & Osaki, Y. (2005). Crystal Modifications and Thermal Behavior of Poly(L-lactic acid) Revealed by Infrared Spectroscopy. *Macromolecules*, 38, 8012-8021.

11) Yasuniwa, M., Tsubakihara, S., Sugimoto, Y. & Nakafuku, C. (2004). Thermal analysis of the double-melting behavior of poly(L-lactic acid). *J Appl Polym Sci Part B : Polym Phys*, 42, 1, 25-32.

12) Yasuniwa, M., Iura, K. & Dan, Y. (2007). Melting behavior of poly(L-lactic acid): Effects of crystallization temperature and time. *Polymer*, 48, 18, 5398-5407.

13) Yasuniwa, M., Sakamo, K., Ono, Y. & Kawahara, W. (2008). Melting behavior of poly(L-lactic acid): X-ray and DSC analyses of the melting process. *Polymer*, 49, 7, 1943-1951.

14) Pilla, S., Gong, S., Neil, E. O., Rowell, R. & Krzysik, A.M. (2008). Polylactide-pine wood flour composites. *Polym Eng Sci*, 48, 3, 578-587.

15) Mathew, A.P., Oksman, K. & Sain, M. (2005). Mechanical properties of biodegradable composites from poly lactic acid (PLA) and microcrystalline cellulose (MCC). *J. Appl Polym Sci*, 97, 5, 2014-2025.

16) Lee, S. (1998). *Thermoplastic Polyurethane Markets in EU: Production, Technology, applications and Trends.* Shawburry, Shrewsbury, Shropshire SY4 4NR, UK: Rapra Technology LTD.

17) Södergård, A. & Stolt, M. (2002). Properties of lactic acid based polymers and their correlation with composition. *Prog Polym Sci*, 27, 6, 1123-1163.

18) Li, B.-H. & Yang, M.-Ch. (2006). Improvement of thermal and mechanical properties of poly(L-lactic acid) with 4,4-methylene diphenyl diisocyanate. *Polym Adv Technol*, 17, 6, 439-443.

19) Raj, R.G., Kokta, B.V., Maldas, D. & Daneault, C. (1989). Use of wood fibers in thermoplastics. VII. The effect of coupling agents in polyethylene–wood fiber composites. *J Appl Polym Sci*, 37, 4, 1089-1103.

20) Woodhams, R. T., Thomas, G. & Rodgers, D. K. (1984). Wood Fibers as Reinforcing Fillers for Polyolefins. *Polymer Eng Sci*, 24, 15, 1166-1171.

21) Demjen, Z., Pukanszky, B. & Nagy, J. (1998). Evaluation of interfacial interaction in polypropylene/surface treated $CaCO_3$ composites. *Compos Part A*, 29, 3, 323–329.

22) Mareri, P., Bastide, S., Binda, N. & Crespy, A. (1998). Mechanical behaviour of polypropylene composites containing mineral filler: Effect of filler surface treatment. *Compos Sci Technol*, 58, 5, 747–752.

23) Li, Y. & Weng, W. (2008). Surface modification of hydroxyapatite by stearic acid: Characterization and in vitro behaviors. *J Mater Sci: Mater Med*, 19, 1, 19–25.

24) Nakagaito, A. N., & Yano, H. (2004). The effect of morphological changes from pulp fiber towards nano-scale fibrillated cellulose on the mechanical properties of high-strength plant fiber based composites. *Appl Phys Mater Sci Process*, 78, 4, 547-552.

25) Lu, J., Wang, T., & Drzal, L. T. (2008). Preparation and properties of microfibrillated cellulose polyvinyl alcohol composite materials. *Compos Part A, 39,* 5, 738-746.

26) Samir, M. A. S. A., Alloin, F., Paillet, M., & Dufresne, A. (2004). Tangling effect in fibrillated cellulose reinforced nanocomposites. *Macromolecules,* 37, 11, 4313-4316.

27) Mamosova, M., Laurova, M. & Nemcokova, V. (2002). *Analysis of Structure of Beech Wood Subjected to Hydrothermal Treatment.* Zvolen, Slovakia: Arbora Publishers.

28) Wikberg, H. & Maunu, S. (2004) Characterisation of thermally modified hard- and softwoods by 13C CPMAS NMR. *Carbohydr Polym,* 58, 4, 461–466.

29) Bhuiyan, T. & Hirai, N. (2001). Effect of intermittent heat treatment on crystallinity in wood cellulose. *J Wood Sci,* 47, 5, 336–341.

30) Boonstra, M. & Tjeerdsma, B. (2006). Chemical analysis of heat treated softwoods. *Holz Roh Werkst,* 63, 3, 204–211.

31) Zanuttini, M., Marzocchi, V. & Citrini, M. (1999). Alkaline treatment of poplar wood. *Holz Roh Werkst;* 57, 3, 185-190.

32) Ichazo, M., N., Abano, C., González, J., Perera, J. & Candal, M., V. (2001). Polypropylene/wood flour composites: treatments and properties. *Compos Struct,* 54, 2-3, 207-214.

33) Swallowe, G.M. (1999). *Mechanical Properties and Testing of Polymers: an A-Z reference.* P.O. Box 17, 3300 AA, Dordrecht, The Netherlands: Kluwer Academic Publishers.

34) Brown, M.E. (2001). *Introduction to Thermal Analysis: Techniques and Applications* (Second edition). P.O. Box 17, 3300 AA, Dordrecht, The Netherlands: Kluwer Academic publishers.

35) Silberschmidt, V.V. (2010). *Computational and Experimental Mechanics of Advanced Materials* (CISM Courses and Lectures, Vol. 514). Udine, Italy: Springer WienNewYork.

36) Chartoff, R. P.; Weissman, P. T. & Sircar, A. (1994). *Assignment of the Glass Transition.* Seyler, R. J.Atlanta, USA: ASTM International.

37) Herzog, B., Gardner, D.J., Lopez-Anido, & R., Goodell, B. (2005). Glass-transition Temperature Based on Dynamic Mechanical Thermal Analysis Techniques as an

Indicator of the Adhesive Performance of Vinyl Ester Resin. *J Appl Polym Sci*, 97, 2221–2229.

38) Höhne, G., Hemminger, W. & Flammersheim, H.-J. (2003). *Differencial Scanning Calorimetry* (Second edition). Germany: Springer-Verlag Berlin Heidelberg New York.

39) Dean, J.A. (1995). *The Analytical Chemistry Handbook*; New York, US: McGraw Hill, Inc.

40) Pungor, E. (1995). *A Practical Guide to Instrumental Analysis*; Boca Raton, Florida: CRC Press.

41) Skoog, D.A., Holler, F.J. & Nieman, T.A. (1998). *Principles of Instrumental Analysis*. (Fith Edition). New York, US: Saunders College Pub.

42) Seydibeyoğlu, M. Ö., & Oksman, K. (2008). Novel nanocomposites based on polyurethane and micro fibrillated cellulose. *Comp Sci Tech*, 68, 3-4, 908-914.

43) Dubief, D., Samain, E., & Dufresne, A. (1999). Polysaccharide microcrystals reinforced amorphous poly(beta-hydroxyoctanoate) nanocomposite materials. *Macromolecules, 32*, 18, 5765-5771.

44) Barham, P.J., Keller, A., Otun, E.L. & Holmes, P.A. (1984). Crystallization and morphology of a bacterial thermoplastic: poly-3-hydroxybutyrate. *J Mater Sci*, 19, 9, 2781–2794.

45) Raya, S. S., Yamada, K., Okamoto, M. & Ueda, K. (2003). New polylactide-layered silicate nanocomposites. 2. Concurrent improvements of material properties, biodegradability and melt rheology. *Polymer*, 44, 3, 857-866.

46) Dubief, D., Samain, E., & Dufresne, A. (1999). Polysaccharide microcrystals reinforced amorphous poly(beta-hydroxyoctanoate) nanocomposite materials. *Macromolecules, 32*, 18, 5765-5771.

47) Nishio, Y., Hirose, N., & Takahashi, T. (1989). Thermal-analysis of cellulose poly(ethylene oxide) blends. *Polym J*, 21, 4, 347-351.

48) John, M.J. & Thomas S. (2008). Biofibres and biocomposites. *Carbohydr Polym*, 71, 343-364.

49) Bengtsson, M., Stark, N. M. & Oksman, K. (2007). Durability and mechanical properties of silane cross-linked wood thermoplastic composites. *Comp Sci Tech*, 67, 2728-2738.

8 Index of tables

Table 1: Sample labelling. 19

Table 2: Percentage changes of storage modulus in rubbery region (80°C) related to the unannealed PVA or PVA/WF of relevant WF concentration. 23

Table 3: Viscoelastic, thermal and tensile properties of PLA, PLA/WF and PLA/MDI composite films with 20% juwenile, mature and compression wood 24

Table 4: Sitka spruce reinforced PLA matrix: Viscoelastic, thermal and tensile properties 28

Table 5: Beech wood flour reinforced PHB matrix – resuplts from DMA, DSC and mechanical testing 29

Table 6: An overview of percentage changes in crystallinity, viscoelastic and tensile properties, and changes of T_g and T_m for PLA and PHB composites due to the chemical treatment of the fillers. The changes are related to composites with unmodified fillers. . 30

Table 7: Results of DMA, DSC and tensile testing. The results are given as percentage changes in composite properties related to the neat material. Elongation at break (ε), tensile strength (TS), Youngs's modulus (E'). T_g and T_m are real changes of temperatures in °C. 33

9 Table of figures

Figure 1: Interphase between fiber and matrix... 9

Figure 2: Phase angle between stress and strain amplitudes within DMA measurement for a) elastic, b) viscous, c) visco-elastic and d) complex modulus diagram for viscoelastic material 13

Figure 3: Example for T_g determination from DMA output demonstrating the production of different values obtained in dependency of method used. Onset of the E', peak of E'' and $tan\ δ$. (Adapted from: Herzog et al. (2005) Glass-transition temperature based on dynamic mechanical thermal analysis techniques as an indicator of the adhesive performance of vinyl ester resin.) 14

Figure 4: The DSC curve typical for thermoplastic polymer. 15

Figure 5: Scheme of the work 18

Figure 6: An example of DMA results: PLA with 20% of softwood flour annealed at 90°C. Improvement of E' in glassy and rubbery region due to WF increasing content. 23

Figure 7: Effectiveness of the filler versus WF content and annealing temperature 23

Figure 8: Scanning electron micrographs of a) nonfibrillated flax , b) flax NFC, c) nonfibrillated wheat straw and d) wheat straw NFC. White scale bars represent the length of 20 μm (a and c) and 3 μm (b and d).. 26

Figure 9: Improvement of PVA viscoelastic properties due to the nFC presence (1 Hz) DMA results of PHB with 20% of modified and unmodified beech wood flour. a) increase in storage modulus due to the beech WF reinforcement and WF modifications, b) decrease and shift of loss tangent due to the reinforcement 26

Figure 10: DSC results: melting temperatures and heat of fusion for PVA/F-NFC and PVA/WS-NFC over the w_c=5-40% NFC mass fraction range 26

Figure 11: Water-uptake of cellulose, PVA and PVA-NFC composites as well as calculated values based on water absorption of neat components. 27

Figure 12: DMA results of PHB with 20% of modified and unmodified beech wood flour. a) increase in storage modulus due to the beech WF reinforcement and WF modifications, b) decrease and shift of loss tangent due to the reinforcement 29

Figure 13: A) storage moduli of PLA/Sitka composites. Amount of sitka JW, MW and CW wood flour was 20%. Apparent improvement of storage modulus in rubbery region; B) $tan\ δ$ of PLA/Sitka composites. Decrease of loss tangent and shift to higher temperature is dependant on the type of WF used 30

10 Selected publications

Publication A

Hrabalova, M., Gregorova, A., Wimmer, R., Sedlarik, V., Machovsky, M. & Mundigler, N.

Effect of wood flour loading and thermal annealing on viscoelastic properties of poly(lactic acid) composite films.

Journal of Applied Polymer Science, 2010, 118, 3, 1534–1540.

Effect of Wood Flour Loading and Thermal Annealing on Viscoelastic Properties of Poly(lactic acid) Composite Films

M. Hrabalova,[1] A. Gregorova,[2] R. Wimmer,[3] V. Sedlarik,[4,5] M. Machovsky,[4] N. Mundigler[1]

[1]*Institute for Natural Materials Technology, Department for Agrobiotechnology, IFA-Tulln, University of Natural Resources and Applied Life Sciences, Vienna, A-1180, Austria*
[2]*Institute for Chemistry and Technology of Materials, Graz University of Technology, Graz 8010, Austria*
[3]*Faculty of Forest Sciences and Forest Ecology, University of Göttingen, Göttingen 37077, Germany*
[4]*Polymer Centre, Faculty of Technology, Tomas Bata University in Zlin, Zlin 76272, Czech Republic*
[5]*Jozef Stefan Institute, Jamova Cesta 39, 1000-Ljubljana, Slovenia*

Received 29 September 2008; accepted 25 March 2010
DOI 10.1002/app.32509
Published online 3 June 2010 in Wiley InterScience (www.interscience.wiley.com).

ABSTRACT: Poly(lactic acid) (PLA) films filled with up to 50 wt % softwood flour were prepared by melt compounding and thermocompression. Thermal annealing of the melt was performed at temperatures from 90°C to 120°C, for 45 min. Responses on polymer-filler interactions, viscoelastic properties, crystallinity of PLA as well as PLA-wood flour-filled films were investigated by differential scanning calorimetry (DSC), dynamic mechanical analysis (DMA), and scanning electron microscopy (SEM). The effectiveness of fillers on the storage moduli (C) was also calculated. The results reveal that wood flour (WF) in conjunction with thermal annealing affected the melting behavior of PLA matrix, and the glass transition temperature. It was further found that the effectiveness of the wood filler in biocomposites widely improved with thermal annealing as well as with higher WF concentration. Finally, it was found that the compatibility between WF and the PLA matrix can be improved when suitable annealing conditions are applied. © 2010 Wiley Periodicals, Inc. J Appl Polym Sci 118: 1534–1540, 2010

Key words: composites; viscoelastic properties; thermal properties; crystallization; annealing

INTRODUCTION

Finding new applications of biodegradable polymeric materials for nonfood commodities and other products is a logical consequence of an increasing environmental awareness. The effort to decrease existing environmental load caused by the annually rising amount of plastic waste has been already considered in legislations.[1] As an example, European regulations for end-of-life vehicle recycling have created interest to look out for materials that are environmentally more compatible and biodegradable.[2] Biodegradable polymers may be divided into three groups: (1) Biopolymers of natural origin, (2) synthetic biodegradable polymers, and (3) modified polymeric materials to reach biodegradability.[3] Poly(lactic acid) (PLA) is a biodegradable, thermoplastic, aliphatic polyester, derived from renewable resources.[4,5] PLA has gained attention as a replacement for conventional synthetic polymeric packaging as well as construction materials during the past decade.[6,7] This biodegradable polymer has several drawbacks limiting its wider use in practice, which is brittleness, low-softening temperature, and the high price.[8] The existing problems have led to efforts to bring in PLA compatible modifiers, additives or fillers that improve mechanical properties, reduce cost of the final product and retain biodegradability at the same time.[9–12] Cellulose-derived fillers seem to meet all discussed demands and their potentialities are intensively investigated.[4,7,8,13–15] The reinforcing effect of the cellulose-based fibers improves mechanical and viscoelastic properties mainly the stiffness[7,11,16] due to the high strength of the fibers. Besides, cellulosic materials were chemically modified to improve the adhesion between the fiber surface and the polymer matrix,[7,12] which is poor due to the counter polarity of the substances. On the other hand, the interfacial performance of PLA-WF

Correspondence to: M. Hrabalova (marta.hrabalova@boku.ac.at).
Contract grant sponsor: Austrian Science Fund (FWF); contract grant number: L319-B16.
Contract grant sponsor: Austrian Ministry of Science and Technology and Ministry of Education, Youth and Sport of the Czech Republic; contract grant numbers: CZ 05/2009, MEB 060908.
Contract grant sponsor: Science and Education Foundation of the Republic of Slovenia (Program "Ad futura").

Journal of Applied Polymer Science, Vol. 118, 1534–1540 (2010)
© 2010 Wiley Periodicals, Inc.

composite is also influenced by the transcrystalline region promoted on the fiber surface.[4,10,17] Generally, the physical constraints originating from the crystalline structure of the PLA, as well as other polymers, have crucial effect on the resulting macroscopic features including mechanical and thermal properties of PLA.[6,18–22] Crystal modifications of isothermally crystallized PLA were described by Zhang et al.[23] and its complex melting behavior by Yasuniwa et al.[24–26] It was proposed that the crystallization mechanism of PLA changes with the crystallization temperature.

On the basis of previously published observations, we hypothesize that the interfacial interaction between PLA and wood flour (WF) might be strongly dependent on the nature of PLA crystalline morphology, particularly on the thermal history of the composite and the number of nucleating sites in PLA. Dynamic mechanical analysis (DMA) was especially used as a highly sensitive method suitable for monitoring polymer-fiber interfaces[27] investigating changes introduced by different crystallization mechanisms as affected by the annealing temperature (T_A) during processing. Thermal and structural properties were also investigated by differential scanning calorimetry (DSC) and scanning electron microscopy (SEM), respectively.

EXPERIMENTAL

Materials

Poly(lactic acid) 7000D obtained from NatureWorks was used as the matrix material. The PLA was reinforced with commercially available softwood flour supplied by J. Rettenmaier & Söhne GmbH + Co., Germany. The softwood flour was ground and homogenized to finally pass a 120 μm sieve.

Sample preparation

Polymer composites based on PLA and WF were prepared by melt mixing in a brabender kneader operating at 190°C and 20 rpm. PLA was first melted in the brabender and then mixed with the WF for 3 min. The composites were molded to thin films with a thickness of ~ 2.50 mm in a hot-press at 160°C preheated for 3 min before a pressure of 10 MPa was applied at 160°C for another 3 min. For thermal annealing, the molded films were kept in the temperature-controlled press for 45 min at a pressure of 10 MPa. Annealing temperatures were 90, 100, 110, and 120°C, respectively. Films were cooled down to room temperature by placing the molds into a cold press. The samples are designated as PLA_X_Y, with X indicating the wood flour concentration (WF_c) in wt %, and Y indicating the annealing temperature

with a 0 for the unannealed samples. Before testing samples were conditioned for 1 week at 23°C and 50% relative humidity (RH).

Differential scanning calorimetry

Thermal analysis was conducted on a Netzsch DSC 200 F3 Maia. Approximately 10 mg of sample was sealed in an aluminum pan. DSC scans were performed at a temperature range between −20°C and 180°C, at 10°C/min heating rate and a nitrogen gas flow of 60 mL/min. Melting temperatures (T_m) were determined from the melting peaks. Specific melting enthalpy (ΔH_m) of the composites is referred to the actual mass fraction (w) of the PLA matrix and was calculated according to (1)

$$\Delta H_m = \frac{\Delta H_{mexp} - \Delta H_c}{w} \quad (1)$$

where ΔH_c is the enthalpy of cold-crystallization, and ΔH_{mexp} the heat of fusion obtained from the melting endotherm.[28]

Dynamic-mechanical properties

Viscoelastic properties of neat PLA and the PLA-WF films included storage modulus (E') as well as the loss factor $\tan \delta = E''/E'$, with E'' being the loss modulus, both determined on a Netzsch DMA 242 C in tensile mode with strips 10 mm × 6 mm × 0.25 mm in size cut from the pressed films. Temperature range was varied between −10°C and +100°C. Measurements were accomplished at a strain sweep frequency of 1 Hz, and a heating rate of 3°C min^{-1}.

Scanning electron microscopy

To visualize the effect of thermal annealing on the PLA-WF composites, i.e., the filler distribution and size, the films were also studied by thermionic-emission scanning electron microscopy (TESCAN VEGA/LMU). The surfaces were prepared by cryogenic fracturing in liquid nitrogen and then coated with a thin layer of Au/Pd. The microscope was operated under high-vacuum mode at an acceleration voltage of 5 kV.

RESULTS AND DISCUSSION

Thermal annealing and melting enthalpy

DSC records of the investigated samples revealed clear effects of thermal annealing as well as WF content on the melting behavior of the tested composites. In case of neat PLA, the expressed increase of ΔH_m as an indicator for polymer crystallinity was

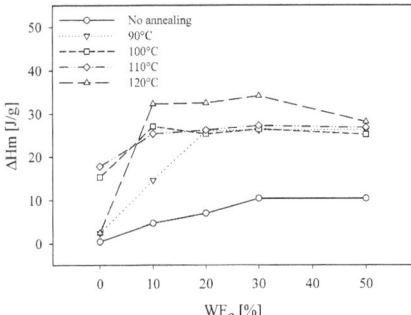

Figure 1 Specific melting enthalpy of neat PLA and PLA-WF composites as related to wood flour content (WF_c) and the annealing temperature.

seen at annealing temperatures around of 100–110°C. The presence of the filler-induced changes in the melting enthalpy of the studied systems. The ability of WF to cause heterogeneous nucleation is proved by the substantial raise of ΔH_m along with increasing WF concentration. A nucleating effect of the WF filler is seen up to 30 wt % in the case of the unannealed composite. A large number of nucleating sites above this concentration has most likely led to a lack of space for spherulites, which impeded further advancement of the crystalline fraction.[29] Annealing had a pronounced influence on the observed ΔH_m. As seen in Figure 1, the higher the annealing temperature, the higher the crystalline fraction present in the system. Melting enthalpy of composites annealed at $T_A = 120°C$ was most expressed. Abe et al.[30] reported discontinuity in crystalline formation at about 120°C. Below this temperature, spherulite growth was evolved, and at higher, temperatures crystal thickening was favored. Thus, it can be proposed that higher ΔH_m might be a consequence of crystal thickening.

The synergism of both the nucleating ability of WF and the annealing temperature has led to highly crystalline material. For example, ΔH_m of PLA_10_120 was almost six times as high as unannealed PLA, both filled with 10% WF. With the addition of 10% WF, ΔH_m greatly responded at annealing temperatures of 90°C and 120°C. The ΔH_m response was less expressed at annealing temperatures of 100°C and 110°C, respectively. Melting enthalpy of the samples having flour fillings over 20 wt % seems not to be affected by thermal annealing. This indicates that the nucleating ability of WF above 20 wt % became ineffective. Consequently, the further development of the crystalline portion was affected by thermal annealing only.

Depending on preparation conditions, the three different crystalline modifications α, β, and γ of PLA are described.[31] The most common crystalline modification in melt-crystallized poly-L-lactic (PLLA) is the orthorhombic α-form and this modification is usually revealed at thermal annealing above 113°C. The crystalline modification formed below this temperature is not yet unambiguously clarified. Authors described the crystalline structure formed to be trigonal β-form;[31] on the other hand, it is also seen as a disordered α-form (pseudoorthorhombic).[24]

The influence of WF content and thermal annealing on melting endotherms is illustrated in Figure 2; thermal characteristics are summarized in Table I. The melting behavior of PLA is complex with regard to its multiple melting behavior as well as polymorphism and was intensively studied by several authors.[24,25,31–33] Yasuniwa et al. described the multiple melting behavior of PLA in dependence on crystallization temperature and stated the discrete change of melting and crystallization behavior (113°C).[31] In this study, a double melting behavior of PLA and PLA-WF, respectively, was observed for the sample sets annealed at $T_A < 110°C$. A low-temperature peak (T_{m_1}) located between 143°C and 149°C, and a high-temperature peak (T_{m_2}) between 151°C and 156°C was detected. The latter can be attributed to the melting of recrystallized crystals[26,33] (Table I). A shift of T_{m_1} toward higher temperatures with higher T_A was seen across the entire WF range. These results are in accordance with findings reported by Masirek et al.[11] For neat PLA films, the observed shift of T_{m_1} might be due to crystalline growth along with T_A,[18,31] which is a consequence of concurrently declining nucleation rates. Therefore, T_{m_1} of PLA can be described as melting of small

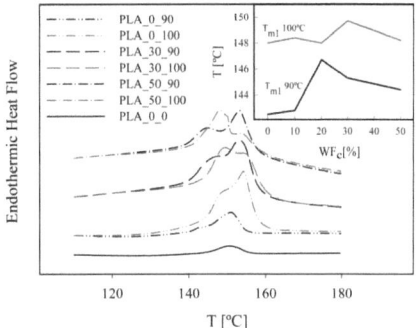

Figure 2 DSC melting endotherms displaying double melting peak detection for $T_C = 90°C$ and $100°C$ and $WF_c = 30$ and 50 wt % positive T_{m_1} shift at higher T_A is seen across the entire WF_c range.

TABLE I
Thermal Characteristics Derived from DSC Endotherms

Sample	ΔH_{m_1} (J/g)	ΔH_{m_2} (J/g)	$\Delta H_{m_1}/\Delta H_{m_2}$	ΔT_{m_1} (°C)	ΔT_{m_2} (°C)
PLA_0_90	0.07 ± 0.01	4.59 ± 0.02	0.015	143	151
PLA_0_100	0.77 ± 0.12	16.19 ± 0.23	0.048	148	154
PLA_10_90	0.32 ± 0.03	18.62 ± 0.10	0.017	143	155
PLA_10_100	2.55 ± 0.11	24.46 ± 0.18	0.104	148	155
PLA_20_90	5.00 ± 0.40	20.96 ± 0.22	0.239	147	156
PLA_20_100	2.63 ± 0.34	22.62 ± 0.37	0.116	148	153
PLA_30_90	5.16 ± 0.30	20.94 ± 0.29	0.246	145	153
PLA_30_100	3.75 ± 0.30	22.60 ± 0.29	0.166	149	155
PLA_50_90	6.34 ± 0.30	19.86 ± 0.68	0.319	144	153
PLA_50_100	9.36 ± 0.45	15.73 ± 0.55	0.595	148	154

[a] Standard deviations were based on triple DSC measurement evaluations.

crystallites having low-thermal stability.[33] For WF-filled samples, a T_{m_1} shift might be the combination of crystalline growth and lamellar thickening of transcrystalline region.[34] For $T_A \geq 110°C$, low-temperature peak T_{m_1} has merged with the T_{m_2} peak at 153–155°C. Melting temperature of the neat PLA was seen at 151°C. The absence of an exothermic peak between T_{m_1} and T_{m_2} indicates that the rate of recrystallization has overwhelmed the rate of melting.[32] As shown in Table I, the area ratio of the low-temperature peak to the high-temperature peak ($\Delta H_{m_1}/\Delta H_{m_2}$) increased with WF_c at each T_A. This contribution for ΔH_m of the low-temperature peak can be explained by a tendency of transcrystalline region development at the PLA-WF interface.[16] Accordingly, the advancement of the area of the low-temperature peak might be caused by the melting at the transcrystalline region, which was markedly extended at higher WF_c.

Viscoelastic properties

Effects of WF content and thermal annealing on the storage moduli (E') in at the glassy and rubbery region (20°C and 80°C, respectively) are listed in Table II. As seen in Figure 3, the capability of the composites to store mechanical energy and resist deformation has increased as WF gets incorporated. This is due to higher rigidity of the filler and the reinforcing effect in the PLA matrix.[35–39] E' in the glassy region was not susceptible to thermal annealing with exception of higher WF loaded samples at 110°C (PLA_30_110 and PLA_50_110). By contrast, the rubbery region of the composites was considerably affected by thermal annealing as a result of the improved thermal stability.[15] The improvement of E' was paralleled with higher WF_c. This effect was expressed by the factor C, which is the effectiveness of the filler on E' in a given polymer matrix:[38]

$$C = \frac{(E'_g/E'_r)_{comp}}{(E'_g/E'_r)_{matrix}} \quad (2)$$

where E'_g and E'_r are storage moduli measured in the glassy and the rubbery regions, at a sweep frequency of 1 Hz. The factor C is inversely proportional to the filler effectivity in the composite. Lower the C, higher the effectiveness of the filler on the composite storage modulus. The effectiveness of the

TABLE II
Variation of Glass Transition Temperature, Loss Factor, and Storage Moduli in Glassy and Rubbery State, with WF_c and Annealing Temperature

	T_g (°C)	tan δ	E' at 20°C (GPa)	E' at 80°C (GPa)
PLA_0_0	63	2.307	2.59	0.025
PLA_0_90	63	2.123	2.157	0.080
PLA_0_100	65	0.709	2.487	0.065
PLA_0_110	63	1.336	2.309	0.019
PLA_0_120	63	1.890	2.345	0.029
PLA_10_0	65	1.989	2.694	0.013
PLA_10_90	64	0.477	2.564	0.138
PLA_10_100	70	0.224	2.598	0.483
PLA_10_110	65	0.217	2.701	0.527
PLA_10_120	67	0.219	2.757	0.599
PLA_20_0	64	1.462	2.654	0.035
PLA_20_90	66	0.236	3.185	0.637
PLA_20_100	68	0.195	2.773	0.618
PLA_20_110	65	0.205	2.818	0.603
PLA_20_120	65	0.180	2.83	0.707
PLA_30_0	64	1.204	3.057	0.053
PLA_30_90	68	0.177	3.661	0.984
PLA_30_100	68	0.164	3.802	1.169
PLA_30_110	65	0.190	4.502	1.008
PLA_30_120	64	0.174	3.087	0.750
PLA_50_0	68	0.631	3.883	0.143
PLA_50_90	67	0.155	4.012	1.328
PLA_50_100	71	0.146	3.855	1.546
PLA_50_110	65	0.137	5.972	1.941
PLA_50_120	66	0.129	3.713	1.364

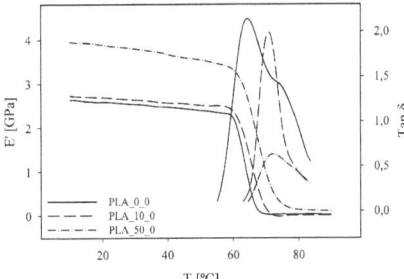

Figure 3 Storage modulus and tan δ as related to temperature measured for unannealed PLA films containing 0, 10, and 50% of WF. Reduced tan δ peaks as well as improved storage moduli are going along with higher WF_c.

filler on E' of nonannealed PLA-WF film was lowest for the system with 10% WF loading and highest for the 50% WF loading. As seen in Figure 4, the C factor dropped with increasing WF_c and T_A, but only up to a T_A of 110°C. Results confirm the thermal stability in the rubbery region, which is caused by WF with reference to thermal annealing. The stress transfer between WF and the matrix weakened at T_A = 120°C (see Fig. 3).

The maximum rate of turndown of the E' was attributed to the glass transition temperature (Table II). Although the T_g of unfilled unannealed samples is not susceptible to thermal annealing, T_g of WF reinforced samples moved to higher temperatures as a consequence of restricted dynamics of polymer chains in confined environments.[39]

Loss factor (tan δ) is an indicator for mechanical damping or internal friction in a viscoelastic system.[38] In composites, a lower tan δ indicates better interfacial bonding between filler and matrix.[27,38] The highest tan δ was measured for neat, unannealed PLA. As seen in Table II, tan δ is inversely proportional to WF_c for a given annealing temperature.[27] The restriction of molecular chain movements is driven by two factors: (1) matrix crystallinity[40] and (2) presence of WF. The crystalline portions in PLA matrices filled with 20–50% wood fibers, annealed at T_A = 90–100°C, turned out to be constant, whereas though tan δ decreased within the same T_A range. This indicates improved interaction between filler and matrix as a consequence of different polymer states in connection with filler particles.[37] It is assumed that transcrystalline growth within annealing temperature range of 90–100°C has improved interfacial bonding between fiber and the matrix.

Scanning electron microscopy

Fracture surfaces of unannealed PLA composites filled with 10% WF [Fig. 5(a)] showed a relatively smooth surface with WF particles having 20 μm in length and about 10 μm in thickness. The effect of thermal annealing is clearly seen in Figure 5(b). The fracture surface appeared fissured, which reveals the higher brittleness of the material. The fracture surface of the composite with the highest investigated WF content, unannealed and annealed, is seen in Figure 5(c) (PLA_50_0) and Figure 5(d) (PLA_50_120), respectively. The highly filled PLA composites are characteristic by a dense and uniform distribution of WF particles within the polymer matrix. The microstructure of the cold fracture of annealed PLA_50_120, which had significantly higher crystallinity than the unannealed comparison (PLA_50_0), shows considerable higher level of WF-PLA matrix isolation. This was most likely caused by the rigidity of the polymer due to organization of the polymeric chains into geometric units. This has led to a loss of the polymer-filler cohesion as seen in the presented SEM pictures. The obtained results are in coherence with the increased C factor, as determined through DMA.

CONCLUSIONS

In this study, PLA—wood fiber composites were subjected to thermal annealing to observe the effects on different properties. Results revealed that WF in conjunction with thermal annealing had strong effects on the melting behavior of PLA matrix, and also on the glass transition temperature. Overall, compatibility between WF and the PLA matrix was improved under suitable annealing conditions. It

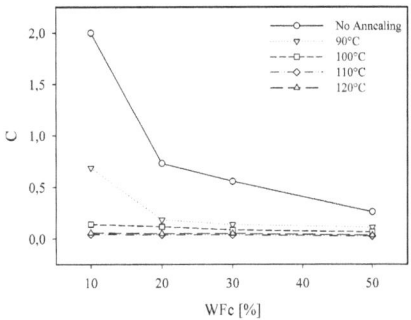

Figure 4 Constant C versus WF_c for different annealing temperatures.

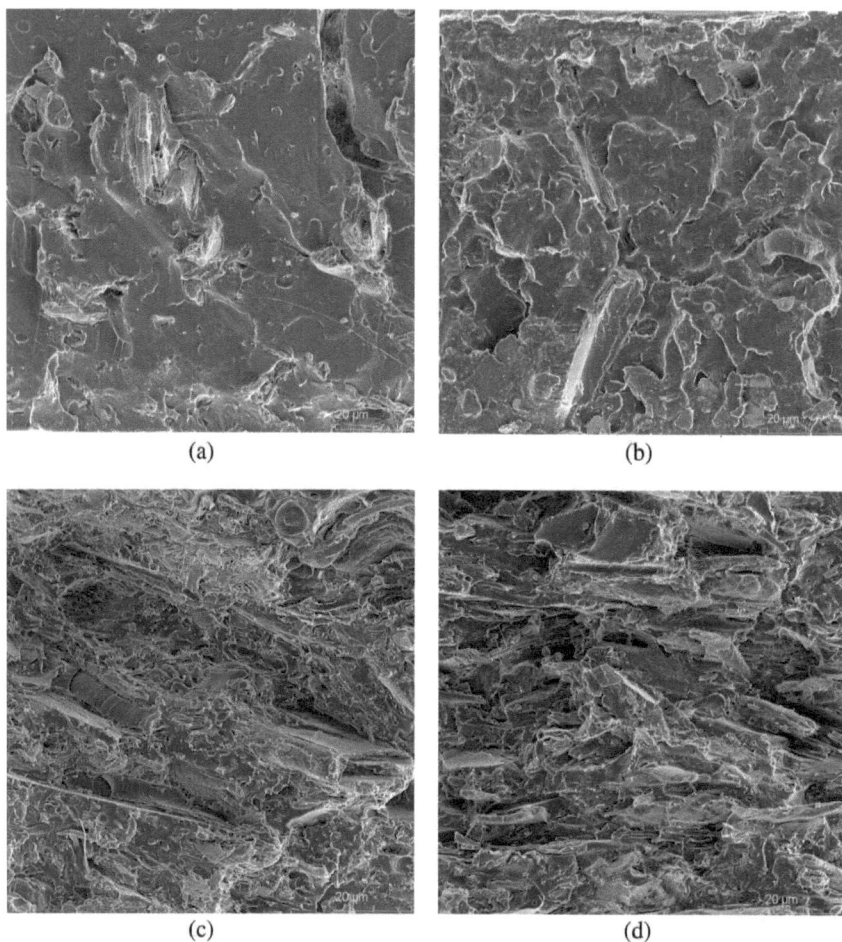

Figure 5 SEM micrographs of fracture surfaces. (a) PLA_10_0 (10% wood fibers, unannealed), (b) PLA_10_120 (10% wood fibers, 120°C annealing, (c) PLA_50_0 (50% wood fibers, unannealled, and (d) PLA_50_120 (50% wood fibers, 120°C annealing). [Color figure can be viewed in the online issue, which is available at www.interscience.wiley.com.]

was found that WF, in conjunction with applied annealing, is both instrumental to influence the crystalline structure of a PLA composite. Therefore, to optimize fiber-matrix compatibility, the crystallization mechanisms need to be better considered in the processing of wood fiber filled PLA composites. With suitable crystalline structure, wood filler content as well as processing conditions, the property profiles of PLA-based composites can be optimized.

References

1. Scott, G. In Degradable Polymers. Principles and Applications; Scott, G., Ed.; Kluwer Academic Publishers: Netherlands, 2002; p 1.
2. Heyde, M. Polym Degrad Stab 1998, 59, 3.
3. Hartmann, M. H. In Biopolymers from Renewable Resources; Kaplan, D. L., Ed.; Springer Academic Press: Berlin, 1998; Vol. 15, p 367.
4. Mathew, A. P.; Oksman, K.; Sain, M. J Appl Polym Sci 2005, 97, 2014.

5. Sedlarik, V.; Saha, N.; Sedlarikova, J.; Saha, P. Macromol Symp 2008, 272, 100.
6. Garlotta, D. J Polym Environ 2001, 9, 63.
7. Huda, M. S.; Drzal, L. T.; Mohanty, A. K.; Misra, M. Compos Sci Technol 2006, 66, 1813.
8. Sedlarik, V.; Saha, N.; Kuritka, I.; Saha, P. J Appl Polym Sci 2007, 106, 1869.
9. Wang, H.; Sun, X.; Seib, P. J Appl Polym Sci 2002, 84, 1257.
10. Pilla, S.; Gong, S.; Neill, E. O.; Rowell, R. M.; Krzysik, A. M. Polym Eng Sci 2008, 48, 578.
11. Masirek, R.; Kulinski, Z.; Chionna, D.; Piorkowska, E.; Pracella, M. J Appl Polym Sci 2007, 105, 255.
12. Gregorova, A.; Hrabalova, M.; Wimmer, R.; Saake, B.; Altaner, C. J Appl Polym Sci 2009, 114, 2616.
13. Sedlarik, V.; Saha, N.; Saha, P. Polym Degrad Stab 2006, 91, 2039.
14. Sedlarik, V.; Saha, N.; Kuritka, I.; Emri, I.; Saha, P. Plast Rubber Compos 2006, 35, 355.
15. Julinova, M.; Kupec, J.; Alexy, P.; Hoffmann, J.; Sedlarik, V.; Vojtek, T.; Chromcakova, J.; Bugaj, P. Polym Degrad Stab 2010, 95, 225.
16. Huda, M. S.; Drzal, L. T.; Mohanty, A. K.; Misra, M. Compos Sci Technol 2008, 68, 424.
17. Mathew, A. P.; Oksman, K.; Sain, M. J Appl Polym Sci 2006, 101, 300.
18. Tsuji, H.; Ikada, Y. Polymer 1995, 36, 2709.
19. Junkar, I.; Cvelbar, U.; Vesel, A.; Hauptman, N.; Mozetic, M. Plasma Process Polym 2009, 6, 667.
20. Junkar, I.; Vesel, A.; Cvelbar, U.; Mozetic, M.; Strnad, S. Vacuum 2009, 84, 83.
21. Vesel, A.; Junkar, I.; Cvelbar, U.; Kovac, J.; Mozetic, M. Surf Interface Anal 2008, 40, 1444.
22. Vesel, A.; Mozetic, M.; Zalar, A. Surf Interface Anal 2008, 40, 661.
23. Zhang, J.; Duan, Y.; Sato, H.; Tsuji, H.; Noda, I.; Yan, S.; Ozaki, Y. Macromolecules 2005, 38, 8012.
24. Yasuniwa, M.; Tsubakihara, S.; Sugimoto, Y.; Nakafuku, C. J Appl Polym Sci Part B: Polym Phys 2004, 42, 25.
25. Yasuniwa, M.; Iura, K.; Dan, Y. Polymer 2007, 48, 5398.
26. Yasuniwa, M.; Sakamo, K.; Ono, Y.; Kawahara, W. Polymer 2008, 49, 1943.
27. Keusch, S.; Haessler, R. Compos A 1999, 30, 997.
28. Sedlarik, V.; Kucharczyk, P.; Kasparkova, V.; Drbohlav, J.; Salkova, A.; Saha, P. J Appl Polym Sci 2010, 116, 1597.
29. Moon, C. K. J Appl Polym Sci 1998, 67, 1191.
30. Abe, H.; Kikkawa, Z.; Inoue, Y.; Doi, Y. Biomacromolecules 2001, 2, 1007.
31. Yasuniwa, M.; Tsubakihara, S.; Iura, K.; Ono, Y.; Dan, Y.; Takahashi, K. Polymer 2006, 47, 7554.
32. Yasuniwa, M.; Tsubakihara, S.; Fujioka, T. Thermochim Acta 2003, 396, 75.
33. Di Lorenzo, M. L. Macromol Symp 2006, 234, 176.
34. Ninomiya, N.; Kato, K.; Fujimori, A.; Masuko, T. Polymer 2007, 48, 4874.
35. Huda, M. S.; Drzal, L. T.; Misra, M.; Mohanty, A. K. J Appl Polym Sci 2006, 102, 4856.
36. Huda, M. S.; Mohanty, A. K.; Misra, M.; Drzal, L. T.; Schut, E. J Mater Sci 2005, 40, 4221.
37. Jamil, M. S.; Ahmad, I.; Abdullah, I. J Polym Res 2006, 4, 315.
38. Pothan, L. A.; Oommen, Z.; Thomas, S. Compos Sci Technol 2003, 63, 283.
39. Zuza, E.; Ugartemendia, J. M.; Lopez, A.; Meaurio, E.; Lejardi, A.; Sarasua, J. R. Polymer 2008, 49, 4427.
40. Chiellini, E.; Covolan, V. L.; Lorenzo, M.; Solaro, E. Macromol Symp 2003, 97, 345.

Publication B

Gregorova, A., Hrabalova, M., Wimmer, R., Saake, B. & Altaner, C.

Poly(lactide acid) Composites Reinforced with Fibers Obtained from Different Tissue Types of Picea sitchensis.

Journal of Applied Polymer Science, 2009, 114, 2616–2623.

Poly(lactide acid) Composites Reinforced with Fibers Obtained from Different Tissue Types of Picea sitchensis

A. Gregorova,[1] M. Hrabalova,[2] R. Wimmer,[3] B. Saake,[4] C. Altaner[5]*

[1]Department of Material Sciences and Process Engineering, Institute of Wood Science and Technology, University of Natural Resources and Applied Life Sciences, Vienna A-1190, Austria
[2]Department of Agrobiotechnology, Institute for Natural Materials Technology, IFA-Tulln, University of Natural Resources and Applied Life Sciences, Tulln A-3430, Austria
[3]Wood Technology and Wood-Based Composites Unit, Faculty of Forest Sciences and Forest Ecology, Georg-August-University Göttingen, Göttingen D-37077, Germany
[4]Johann Heinrich von Thünen-Institut, Federal Research Institute for Rural Areas, Forestry and Fisheries, Institute of Wood Technology and Wood Biology, Hamburg 21031, Germany
[5]Department of Chemistry, University of Glasgow, Glasgow, United Kingdom

Received 18 March 2009; accepted 22 May 2009
DOI 10.1002/app.30819
Published online 7 July 2009 in Wiley InterScience (www.interscience.wiley.com).

ABSTRACT: Wood fibers vary in their properties across species, across trees of the same species, and within single trees. This work takes advantage of wood fibers reinforcing poly(lactic acid) composites that originate from different tissue types of the species Sitka spruce (Picea sitchensis). Fibers were prepared with high temperature thermo-mechanical processing (TMP) from juvenile, mature, and compression wood tissues of Sitka spruce. Composites were made by solution casting with subsequent hot-pressing. Thermal as well as mechanical properties were determined using differential scanning calorimetry (DSC), dynamic mechanical analysis (DMA), and tensile testing. The obtained results showed that the chemical and physical properties of different tissue-type Sitka spruce fibers have significant effects on the thermal and mechanical properties of the Polylactic acid (PLA)/Sitka fiber composites. To increase interfacial compatibility between the hydrophilic fibers and the hydrophobic polymer matrix, the fibers were treated with vinyltrimethoxysilane (VTMO), while PLA was modified with 4,4-methylene diphenyl diisocyanate (MDI). It was found that PLA/Sitka composites treated with VTMO and MDI exhibited improved thermal and mechanical properties, compared to the unmodified control. The work also demonstrates that there is potential to improve biobased composites by utilizing the natural variability of wood fibers. © 2009 Wiley Periodicals, Inc. J Appl Polym Sci 114: 2616–2623, 2009

Key words: biopolymers; composites; mechanical properties; thermal properties; Sitka spruce [Picea sitchensis (Bong.) Carrière]

INTRODUCTION

During recent years, intensive research has been focusing on the development of biobased and biodegradable plastics with the intention to reduce environmental pollution and to replace petroleum-based plastics. Polylactic acid (PLA) is biodegradable hydrolysable aliphatic semi-crystalline polyester produced through direct condensation of its monomer, lactic acid, followed by a ring opening polymerization of the cyclic lactide dimmer. Lactic acid can be obtained from renewable resources such as saccharide-based materials.[1-3] PLA shows stiffness and strength properties comparable to petroleum-based plastics and can be processed by standard methods such as extrusion, injection molding, thermoforming, or compression molding.[4] Further, PLA is a readily compostable and degradable thermoplastic polymer. Despite these promising properties, its applicability is restricted by high production costs, brittleness, and a low softening temperature.[5] Brittleness can be lowered through incorporation of plasticizers,[6] while production costs can be reduced, and the mechanical performance of the material modified by the addition of various fillers. The use of renewable and biodegradable fillers such as starch, cellulose, kenaf, hemp, and wood fibers has been investigated intensively during the past years.[5,7-12]

Wood fibers attracted attention as fillers mainly in polyolefins in the first place because of their

*Present address: School of Biological Sciences, Faculty of Science, University of Auckland, Auckland 92019, New Zealand
Correspondence to: A. Gregorova (adriana.gregorova@boku.ac.at).
Contract grant sponsor: Austrian Science Fund FWF; contract grant number: L319-B16.
Contract grant sponsor: SHEFC (Scottish Higher Education Funding Council).

Journal of Applied Polymer Science, Vol. 114, 2616–2623 (2009)
© 2009 Wiley Periodicals, Inc.

abundance, renewability, nonabrasiveness, low density, and low price. The use of wood fibers lowers production costs and has the potential to positively modify mechanical properties. However, there are drawbacks such as the high variability of the physical fiber properties as well as their low compatibility with hydrophobic thermoplastic polymers. Numerous coupling agents have been investigated to increase the interfacial compatibility between the hydrophilic fillers and the hydrophobic polymer matrix.[6,13–16] Pilla et al.[16] reported improved mechanical properties of PLA-recycled wood fiber composites with the fibers were treated with 0.5 wt % silane. Alternatively, methylenediphenyl diisocyanate (MDI) was successfully as coupling agent to improve the interfacial interaction between the hydrophobic PLA matrix and the hydrophilic filler.[17–20] In addition, the increase of PLA hydrophilicity by chemical modification may promote biodegradation rate.[21]

As chemical modification trials of wood fibers have been frequently applied, no research was done so far on utilizing the natural variability of wood fibers for biocomposites. There are published data on the influence of different wood species on the physical properties of plastic (PVC)/wood-flour composites has been published.[22,23] Major factors influencing the physical properties of composite were reported to be the surface roughness, surface chemistry, and particle size. Although statistical significant differences were found between composites made from wood flours of different species, these influence was relatively small.

Wood fibers vary in their properties across species, across trees of the same species, and within single trees. Variations of the fiber properties with cambial age are well documented.[24–27] This knowledge is exploited in the pulp and paper industry where top logs (juvenile wood) and chips from saw milling (mature wood) are separated to adjust paper quality. Juvenile wood develops in early stages of tree growth. Cells in juvenile wood are shorter and have a smaller cell diameter as well as larger microfibril angles (up to 55) as compared to mature wood. Juvenile wood has a lower density and strength than mature wood; it contains less cellulose, more hemicelluloses (in particular arabinoxylan), and also more lignin. Trees with normal wood grow form juvenile and mature wood, because it is purely a matter of the age. Reaction wood, as another type of wood tissue, is formed when a tree is forced out of its normal, straight growth to compensate for the abnormal growing conditions.[28,29] In softwoods, irregular cells develop on the underside of a stem or branch and are referred to as compression wood. This special type of reaction wood contains more lignin, less cellulose, and less galactoglucomannan compared to normal wood. Furthermore, b-1,3-linked glucan (callose) and b-1,4-linked galactan are also present in compression wood. The microfibril angle in the modified secondary cell wall layer of compression wood is high (> 45). The rounded and thick-walled compression wood fibers (tracheids) are 10–40% shorter than normal fibers. Compression wood is less stiff than normal wood and exhibits a higher strain at breakage.

The objective of this work was to assess the properties of PLA composites reinforced with fibers obtained from different wood tissue types of a Sitka spruce [lat. Picea sitchensis (Bong.) Carrière]. We hypothesize that (1) there is potential to optimize composites properties by utilizing the natural variability of fibers that originate from Sitka spruce, and (2) there is additional potential to improve property profiles through interaction between wood fibers coming from different tissue types and chemical fiber modification, i.e., silane and 4,4-methylene diphenyl diisocyanate treatment. There are no previously published reports on the effect of different wood fiber types of one tree species used with PLA composites.

EXPERIMENTAL PART

Materials

PLA 7000D pellets received from NatureWorks LLC (Minnetonka, MN), were used as matrix material. The PLA had a density of 1.24 g/cm^3. Wood fibers were prepared from a 36-year-old Sitka spruce [P. sitchensis (Bong.) Carrière] tree grown at Kershope (Northumbria, UK). Juvenile wood (JW), mature wood (MW), and severe compression wood (CW) were identified and isolated from the tree. The material was chipped using a hydraulic guillotine. Chips were defibrillated under mild conditions after saturated steam cooking for 4 min with 0.32 MPa pressure at 135.7 C, followed by 30 s mechanical defibrillation. Pulps were then refined at four refiner-plate distances (2 mm, 0.8 mm, 0.3 mm, 0.12 mm) and sorted through a 0.15 mm slit sieve. Centrifuged pulps at 30–40% moisture content were stored frozen until further processing. Splinter content was highest in JW (Table I). Vinyltrimethoxysilane (VTMO) used for fiber modification was obtained from Fluorochem (Derbyshire, UK), and 4,4-methylene diphenyl diisocyanate (MDI) was obtained from Sigma-Aldrich (Germany).

Fiber characterization

Extractive content was determined according to Tappi standard T264 om-88. The carbohydrate composition of the pulps was determined after Sinner et al.[30] and Sinner and Puls[31] applying a two-step hydrolysis with H_2SO_4. Subsequently, the monosaccharides were determined by borate complex anion exchange chromatography. The results were not

TABLE I
Designation of Sitka Spruce Fibers

Sitka spruce fibers	Source of fiber	Cambial age (Years)	Splinter content (%)
JW	Juvenile wood	< 9	25.6
MW	Mature wood	> 20	9.3
CW	Compression wood	10–15	18.0

corrected for sugar losses and water addition during hydrolysis. The hydrolysis residue was gravimetrically determined and denoted as lignin content analogous to the Klason method.

The fiber length was determined on a ZEISS Axioplan 2 optical microscope using the publicly available Scion Image software.[32] Averaged values were based on measurements of 50 complete, unbeaten fibers. Average microfibril angles (MFA) of JW, MW, and CW were measured on a representative radial profile (2 mm tangentially) by wide-angle X-ray scattering (WAXS). The diffraction patterns were obtained on a Rigaku R-AXIS RAPID image-plate diffractometer equipped with Cu radiation using a 0.8 mm collimator. Measurements were done at ambient temperature and relative humidity with the radiation passing through the sample in tangential direction. MFA was calculated according to the variance approach.[33]

Color of the Sitka spruce fibers was characterized using a Konica Minolta colorimeter calibrated to a white standard. The color parameters L*, a*, and b* were determined by the CIELAB method,[34,35] where L* indicates lightness, and a* and b* the chromatic coordinates on green-red and blue-yellow axes, respectively. Additionally, angle (h) and color saturation (C*) were determined.

Fiber treatment

The mass of 0.3 g Sitka spruce fibers were stirred in distilled water for 8 h at room temperature. Fibers were sequentially dehydrated in 15 mL water-methanol (40/60 w/w) followed by pure methanol for 8 h each. For the following silane treatment, the 0.3 g of Sitka spruce fibers were stirred in 15 mL methanol with 0.5 wt % vinyltrimethoxysilane (weight percentage relative to the fiber dry mass) for 4 h. A few drops of acetic acid were added to adjust the mixture to pH 4. Weight percentage gain (WPG) was calculated to indicate the degree of chemical modification of the Sitka spruce fibers:

$$WPG = \frac{W_2 - W_1}{W_1} \cdot 100\% \quad (1)$$

where, W_1 is the weight of the dry sample prior to modification, and W_2 the post modification weight.

Preparation of PLA/Sitka composites

Neat PLA and PLA/MDI were processed according to Li and Yang to prepare PLA/Sitka composites (detailed compositions are listed in Table II).[21] After complete dissolution of 1.2 g PLA in 50 mL chloroform, 0.3 g of wood fibers suspended in 15 mL methanol were added and stirred for 2 h. Mixtures were then casted on Teflon dishes and dried at room temperature for 1 day followed by 2 days at 40 C in vacuum. The obtained films were disintegrated into small flakes and hot pressed at 160 C for 2 min at a pressure of 10 MPa. Films were stored for 3 weeks at standard conditions (23 C, 50% RH) prior to testing.

Differential scanning calorimetry

Thermal behavior of PLA and PLA/Sitka composites was characterized by differential scanning calorimeter (DSC 200 F3, Netzsch). Indium was used for calibration. Data were collected by heating from 30 to 200 C at a constant heating rate of 10 C/min under a constant nitrogen flow of 60 mL/min. Samples with a weight of 10 mg were placed in aluminum pans. Values for glass transition temperatures, fusion enthalpies, and melting temperatures were

TABLE II
Composition of the PLA/Sitka Fiber Composites

Sample	PLA (wt %)	MDI/PLA in ratio 0.25% (wt %)	Sitka fibers (wt %)			Silane treatment of fibers
			JW	MW	CW	
PLA	100	–	–	–	–	–
PLA/JW	80	–	20	–	–	–
PLA/MW	80	–	–	20	–	–
PLA/CW	80	–	–	–	20	–
PLA/JW_silane	80	–	20	–	–	Yes
PLA/MW_silane	80	–	–	20	–	Yes
PLA/CW_silane	80	–	–	–	20	Yes
PLA_MDI/JW	–	80	20	–	–	–
PLA_MDI/MW	–	80	–	20	–	–
PLA_MDI/CW	–	80	–	–	20	–

TABLE III
Physical Properties and Color Parameters of Sitka Spruce Fibers

Sitka fibers	Average fiber length (mm)	Density (kg/dm³)	Microfibril angle ()	L*	a	b	C*	h
JW	2.1	0.361 (0.025)	17.8 (1.3)	67.86	2.73	14.71	13.98	78.73
MW	2.4	0.550 (0.075)	12.9 (0.2)	68.77	3.43	16.07	16.44	77.95
CW	1.7	0.475 (0.039)	33.1 (4.7)	62.04	6.66	22.59	23.55	73.58

Standard deviation in parentheses.

evaluated. Crystallinity was estimated following the equation,

$$Xc\% = \frac{DH_m}{DH_m^0} \cdot \frac{100}{w} \quad (2)$$

where, DH_m^0 is the enthalpy of melting for 100% crystalline PLA being equal to 93.7 J/g,[36] DH_m is the enthalpy for melting of the measured sample, and w is the PLA weight fraction of the measured sample.

Tensile testing

Tensile strength, elongation at break, and Young's modulus of the samples were determined on a 100 N Zwick, Type BZ1, universal testing machine. The rectangular films were 10 mm wide and a 25 mm grip clearance was used. The crosshead speed was 2 mm/min. All mechanical parameters were derived by averaging five experimental runs.

Gel content

The gel content of the composites was determined indirectly through xylene extraction at 110 C for 12 h. After xylene extraction, the samples were filtered, dried to constant weight, and re-weighed. The gel content (GC) of the samples was determined and calculated according to the following equation,

$$GC\% = 100 - \left(\frac{W_1 - W_2}{W_1} \cdot \frac{100}{w}\right) \quad (3)$$

where, W_1 is the weight of the dry sample prior to xylene extraction, W_2 the dry sample weight after xylene extraction, and w is weight fraction of PLA in the sample.

Dynamic mechanical properties

The viscoelastic properties of neat PLA and the PLA/Sitka films including the storage modulus (E^0) as well as the mechanical loss factor (Tand = E^{00}/E^0) were determined using the dynamic mechanical analysis (DMA; 242 C, Netzsch) in tensile mode with strips having 10 × 6 × 0.18 mm cut from the films. Temperature ranged from −10 C to +100 C,

and oscillation frequency was kept at 1 Hz; the heating rate was 3 C/min.

RESULTS AND DISCUSSION

Sitka spruce fiber properties

Compression wood (CW) was easily defibrillated under the used mild thermo-mechanical pulping conditions. This was not the case for juvenile wood (JW), which did not readily imbibe. Reduced permeability of Sitka spruce heartwood, which contains also juvenile wood, is well documented.[37,38] Splinter content after refining for mature wood (MW), CW and JW was 9.3%, 18%, and 25.6%, respectively (Table I). Physical properties of Sitka fibers and their color parameters are shown in Table III. Sitka spruce fibers originating from MW exhibited the highest density and the smallest microfibril angle (MFA) and the longest fibers. It is known that a larger MFA is essential for the pliability of young trees; MFA is also higher in CW. MFA in compression wood was over 33, which was twice the value measured in MW. Fibers differed also with respect to color (Table III). Fibers originating from MW and JW woods were brighter and more yellow, while fibers from CW were darker and reddish. The chemical composition of the Sitka spruce fibers is presented in Table IV. Fibers originating from CW had the highest

TABLE IV
Chemical Composition of Sitka Spruce Fibers

Sitka spruce fiber type			
Chemical content (%)	JW	MW	CW
Extractives[a]	3.1	2.0	2.8
Lignin[b]	28.2	26.6	37.8
Xylose[b]	4.9	4.6	4.6
Glucose[b]	49.9	53.9	39.7
Mannose[b]	12.1	12.6	7.0
Galactose[b]	1.2	1.0	6.5
Arabinose[b]	0.7	0.7	0.5
Ramnose[b]	0.1	–	–
4-O-Methyl[b]	0.6	0.5	0.5

[a] Percentage based on weight of the unextracted Sitka fibers.
[b] Percentage based on weight of the extractive-free Sitka fibers.

lignin and galactose content in conjunction with the lowest glucose, mannose, and arabinose content. Data for CW were in accordance to literature.[29] Differences in chemical composition between JW and MW fibers were less pronounced, though significant. JW fibers exhibited higher xylose and lignin contents compared to those from MW, which was also in accordance to literature.[27,29,39]

Vinyltrimethoxysilane (VTMO) has been used to modify the hydrophilic character of the Sitka spruce fibers. Weight percentage gain (WPG) after silane treatment of the Sitka spruce fibers was 12.01% for JW fibers, 13.55% for MW fibers, and 7.46% for CW fibers. Lower WPG of CW in comparison with those of JW and MW might be explained by the fact that CW has thicker cell walls and fewer pits, which decreased accessibility and surface area and therefore reduced uptake of the silane coupling agent. The lower polysaccharide content of CW cell walls has also resulted in a lower amount of hydroxyl groups which allowed less silicate ester bonds to be available for the silane coupling agent.[40,41]

Thermal properties

DSC heating thermograms of PLA and PLA/Sitka composites recorded between 30 and 200 C are seen in Figures 1–3. Thermal properties such as glass transition temperature (T_g), melting enthalpy (ΔH_m), melting temperature (T_m), and degree of crystallinity (X_c) obtained from the DSC analysis are summarized in Table V. Pure PLA was characterized by a T_g of 46 C, a T_m of 150 C, and a crystallinity X_c of 18.2%. The 20 wt % addition of untreated Sitka spruce fibers raised T_g to 52–54 C, and X_c to 25.0–28.7% depending

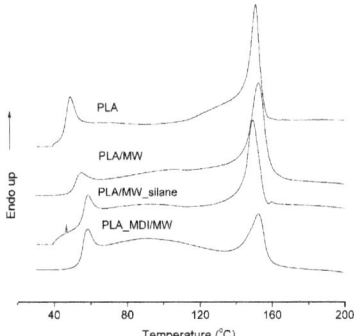

Figure 2 DSC heating thermograms for neat PLA and PLA/Sitka fiber composites made from mature wood (20 wt % fiber content).

on the fiber type (JW, MW, CW). T_m did not change. The shift to a higher T_g could be explained by the restricted mobility of PLA chains due to the presence of reinforcing wood fibers and the increased crystallinity of the PLA. The observation that Sitka spruce fibers acted as nucleating agent and therefore has raised the crystallinity of the PLA matrix is in accordance with literature.[16,42] The PLA/JW and PLA/CW composites exhibited a slightly higher crystallinity than the PLA/MW composite. The surface characteristics of the fibers have most likely influenced their capability of acting as a nucleation point for the crystallization of the PLA matrix. The precipitation of hydrophobic extractives onto TMP fiber surfaces during processing has been reported.[43]

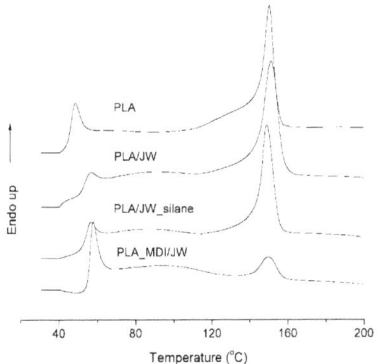

Figure 1 DSC heating thermograms for neat PLA and PLA/Sitka fiber composites made from juvenile wood (20 wt % fiber content).

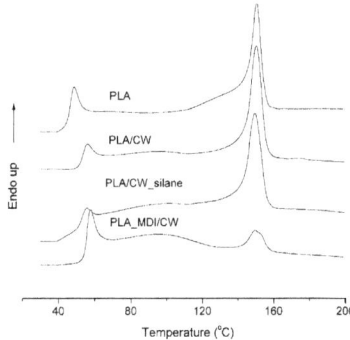

Figure 3 DSC heating thermograms for neat PLA and PLA/Sitka fiber composites made from compression wood (20 wt % fiber content).

TABLE V
Thermal Properties of Neat PLA and PLA/Sitka Fiber Composites

Sample	T_g (C)	DH_m (J/g)	T_m (C)	X_c (%)
PLA	46	17.1	150	18.2
PLA/JW	54	21.5	151	28.7
PLA/MW	52	18.7	150	25
PLA/CW	53	20.9	150	27.7
PLA/JW_silane	54	17.3	149	23.1
PLA/MW_silane	56	14.8	149	19.7
PLA/CW_silane	53	19.0	150	25.4
PLA_MDI/JW	55	4.4	150	6.0
PLA_MDI/MW	55	10.3	152	13.8
PLA_MDI/CW	55	4.8	150	6.4

TABLE VI
Mechanical Properties of Neat PLA and PLA/Sitka Fiber Composites

Sample	Tensile strength (MPa)	Elongation at break (%)	Young's modulus (MPa)
PLA	45.8 (4.7)	2.5 (0.7)	2620 (166)
PLA/JW	35.2 (4.0)	1.2 (0.2)	2930 (398)
PLA/MW	45.3 (7.3)	1.1 (0.2)	3390 (242)
PLA/CW	42.0 (5.2)	1.6 (0.2)	2920 (262)
PLA/JW_silane	38.0 (3.1)	0.9 (0.1)	3120 (405)
PLA/MW_silane	51.5 (6.6)	1.7 (0.1)	3940 (300)
PLA/CW_silane	47.8 (7.5)	1.4 (0.3)	3160 (437)
PLA_MDI/JW	39.6 (2.8)	1.4 (0.2)	2780 (220)
PLA_MDI/MW	55.0 (3.9)	1.3 (0.2)	3500 (230)
PLA_MDI/CW	46.2 (4.2)	1.5 (0.1)	3100 (230)

Standard deviation in parentheses.

From the data in Table V, it is evident that silane treatment of fibers slightly decreased the resulting crystallinity compared to those composites filled with untreated Sitka spruce fibers. This decrease in crystallinity was in agreement with literature, which reported that silane treatment of wood-flour increased the degree of crosslinking in HDPE/wood-flour composites. Crystallinity in turn was decreased.[44] The extent by which the crystallinity of the matrix decreased seemed to correlate with the WPG of silane of the individual fibers. Compared to the untreated PLA/Sitka composites, the silane treatment increased T_g only when MW fibers were used. No shift of T_g due to a silane treatment was recorded for JW and CW fibers. However, for better understanding, we determined glass transition temperature also with DMA since for semi-crystalline polymers higher sensitivity is reported.[45]

The MDI modified PLA/Sitka composites exhibited glass temperatures comparable to those treated with silane, although crystallinity was much lower. This could be attributed to a more branched molecular structures formed in PLA when mixed with MDI.[46] Investigation conducted by Li and Yang suggested that MDI as chain extender may disrupt PLA crystallization.[21]

Mechanical properties

Incorporation of 20 wt % untreated fibers into the PLA matrix slightly decreased tensile strength as well as elongation at break (Table VI). This indicated a low interfacial compatibility between the fibers and the PLA matrix. On the other hand, stiffness of PLA/Sitka composites was improved. The mechanical behavior of the Sitka composites depended on the fiber origin which can be explained by the present chemical differences and consequently different interfacial compatibility between the fibers and polymer matrix. The recorded different behavior of Sitka fibers was promoted by treatment with coupling agents. MW fibers were the most responsive to a silane or MDI treatment, possibly because of the differences in surface chemistry. Huda et al.[47] recorded an improvement of the mechanical properties of PLA/kenaf composites after alkali treatment of the fibers. It is well established that both silane and MDI react with the hydroxyl groups that are present in PLA as well as in the wood fibers.[48]

The presence of a network in the matrix structure was measured indirectly as gel content after extraction in hot xylene. It is believed that the crosslinked part of the composite is insoluble in boiling xylene, while the non-crosslinked part is soluble. Figure 4 shows the determined gel content in PLA/Sitka composites. Composites treated with MDI and filled with fibers from MW exhibited the most extended network structure (32.1% gel content) what corresponded to the best mechanical properties.

Dynamic mechanical properties

Storage modulus and Tand dependent on temperature for neat PLA and untreated PLA/Sitka

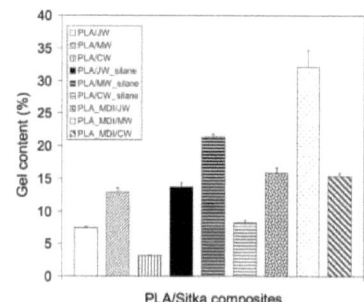

Figure 4 Gel content in PLA/Sitka composites after xylene extraction.

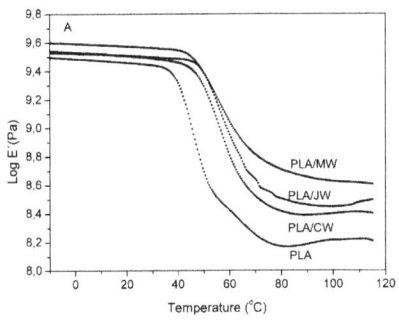

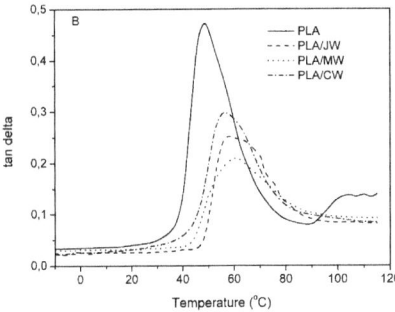

Figure 5 DMA curves for PLA and PLA/Sitka fiber composites containing 20 wt % of untreated fibers. (A) Dependence of storage modulus on temperature and (B) Dependence of Tand on temperature [juvenile wood (JW), mature wood (MW), compression wood (CW)].

composites are shown in Figure 5. The different fiber types improved stiffness of the PLA composites across the measured temperature range. Storage modulus of neat PLA dropped sharply at 38.3 C.

The incorporation of 20 wt % of untreated Sitka spruce fibers shifted this storage modulus drop by 10 C towards higher temperatures. Compared to neat PLA, the Tand peaks of the PLA/Sitka composites have decreased and they also shifted to higher temperatures. The intensity of the Tand peaks of PLA/Sitka fiber composites decreased and shifted to higher temperatures compared to neat PLA. This could be explained by a change in the molecular mobility of PLA molecules due to the incorporation of fibers. Table VII presents DMA data for PLA and PLA/Sitka composites at temperatures of 20, 40, 60, and 80 C, respectively. The storage modulus and T_g (deduced from the Tand peak temperature) increased after silane treatment of the fibers as well as after addition of MDI to the PLA matrix. Incorporation of untreated Sitka spruce fibers increased the storage modulus (stiffness) of the composites by 4–30% at 20 C depending on fibers origin. Fibers from MW embedded into a PLA/MDI matrix increased the storage modulus by 64% at 20 C. This indicated a significant improvement of the interfacial compatibility by the coupling agent. PLA_MDI/Sitka composites showed higher and broader Tand peaks compared to neat PLA, which suggested better damping of these samples. The increase of Tand was probably caused by a formation of branched molecular structures in the PLA/MDI matrix, which was also indicated by the more extended network structure (gel content).[49]

CONCLUSIONS

The thermal, mechanical, and dynamic mechanical properties of Sitka spruce fiber reinforced PLA composites have been investigated. Fibers originated from mature, juvenile, and compression wood of Sitka spruce.

Hypothesis 1 stating a potential for optimizing composite properties by utilizing the natural

TABLE VII
DMA Properties of Neat PLA and PLA/Sitka Fiber Composites at Different Temperatures

Sample	Storage modulus E' (MPa)					Tand
	20 C	40 C	60 C	80 C	(C)	Max. Intensity
PLA	2876	2175	270	147	48.4	0.474
PLA/JW	3018	2822	719	402	57	0.249
PLA/MW	3736	3504	1107	511	60.7	0.21
PLA/CW	3209	2880	641	261	56	0.297
PLA/JW_silane	3332	3061	1158	347	62.9	0.251
PLA/MW_silane	3928	3744	1704	615	62.1	0.201
PLA/CW_silane	3503	3197	1081	430	61.5	0.205
PLA_MDI/JW	3447	3281	473	123	59.1	0.526
PLA_MDI/MW	4720	4400	2106	663	65.8	0.621
PLA_MDI/CW	3853	3655	509	175	60	0.503

Oscillation frequency 1 Hz.

variability of fibers present in a single tree species (here P. sitchensis) was accepted. It was shown that mature fibers improved Young's modulus of the tested PLA composites by 30%, while juvenile fibers did so by 12% only. In contrast to the other fiber tissue types, who showed a decline, mature wood fibers kept the tensile strength of PLA/MW composites unchanged in comparison to neat PLA. The most effective reinforcing fibers originated from mature wood. They exhibited the highest content of glucose and lowest concentration of lignin, which could had an influence on better interfacial interaction with PLA matrix. However, physical characteristics such as high fiber length and surface area should not be ignored.

Hypothesis 2 claiming that fiber modification interacts with fiber type selection was also accepted. It was shown for Young's modulus that with silane treatment an improvement by 50% was possible. For tensile strength MDI treatment combined with mature wood fibers showed a 20% improvement, while silane treatment resulted in a 12% improvement only. The use of silane and isocyanate coupling agents had a positive influence on the thermal and mechanical properties of the composites. This is believed to be due to the improved interfacial compatibility between fibers and PLA matrix.

The variability of the fiber properties needs to be considered when wood fibers are to be used as reinforcing phase in biocomposite materials. The variability might even bear the potential of design materials with tailor-made properties.

The authors thank B. Gardiner and S. Mochan from Forest Research UK, for the provision of the timber.

References

1. Södergard, A.; Stolt, M. Prog Polym Sci 2002, 27, 1123.
2. Vink, E. T.; Rabago, K. R.; Glassner, D. A.; Gruber, P. R. Polym Degrab Stab 2003, 80, 403.
3. Sedlarik, V.; Saha, N.; Kuritka, I.; Saha, P. J Appl Polym Sci 2007, 106, 1869.
4. Garlotta, D. J Polym Environ 2001, 9, 63.
5. Huda, M. S.; Drzal, L. T.; Mohanty, A. K.; Misra, M. Compos Sci Technol 2006, 66, 1813.
6. Martin, O.; Avérous, L. Polymer 2001, 42, 6209.
7. Martin, O.; Schwach, E.; Averous, L.; Couturier, Y. Starch 2001, 53, 372.
8. Oksman, K.; Skrifvars, M.; Selin, J.-F. Compos Sci Technol 2003, 63, 1317.
9. Mathew, A. P.; Oksman, K.; Sain, M. J Appl Polym Sci 2005, 97, 2014.
10. Masirek, R.; Kulinski, Z.; Chionna, D.; Piorkowska, E.; Pracella, M. J Appl Polym Sci 2007, 105, 255.
11. Ikeda, K.; Takatani, M.; Sakamoto, K.; Okamoto, T. Holzforschung 2008, 62, 154.
12. Sykacek, E.; Schlager, W.; Mundigler, N. Polym Compos, published online in Wiley Interscience (www.interscience.wiley.com), 2009.
13. Wu, C.-S. J Appl Polym Sci 2004, 94, 1000.
14. Li, T.; Yan, N. Compos Part A 2007, 38, 1.
15. Zhang, J.-F.; Sun, X. Macromol Biosci 2004, 4, 1053.
16. Pilla, S.; Gong, S. S.; O'neill, E.; Rowell, R. M.; Krzysik, A. M. Nineth International Conference on Wood & Biofiber Plastic Composites; Madison, Wisconsin, May 21-23, 2007, p 376.
17. Wang, H.; Sun, X.; Seib, P. J Appl Polym Sci 2002, 84, 1257.
18. Yu, L.; Petinakis, S.; Dean, K.; Bilyk, A.; Wu, D. Macromol Symp 2007, 249, 535.
19. Jun, Ch. L. J Polym Environ 2000, 8, 33.
20. Wang, H.; Sun, X.; Seib, P. J Polym Environ 2002, 10, 133.
21. Li, B.-H.; Yang, M.-Ch. Polym Adv Technol 2006, 17, 439.
22. Xu, Y.; Wu, Q.; Lei, Y.; Yao, F.; Zhang, Q. J Polym Environ 2008, 16, 250.
23. Kim, J. W.; Harper, D. P.; Taylor, A. M. J Appl Polym Sci 2009, 112, 1378.
24. Macdonald, E.; Hubert, J. Forestry 2002, 75, 107.
25. Mitchell, M. D.; Denne, M. P. Forestry 1997, 70, 47.
26. Uprichard, J. M.; Lloyd, J. A. N Z J Forest Sci 1980, 10, 551.
27. Bertraud, F.; Holmbom, B. Wood Sci Technol 2004, 38, 245.
28. Panshin, A. J.; Zeeuw, C. de. In Textbook of Wood Technology: Structure, Identification, Properties, and Uses of the Commercial Woods of the United States and Canada, 4th Ed.; McGraw Hill: NY, 1980; p 722.
29. Timell, T. E. Compression Wood in Gymnosperms; Springer-Verlag: Heidelberg, New York, p 2150.
30. Sinner, M.; Simatupang, M. H.; Dietrichs, H. H. Wood Sci Technol 1975, 9, 307.
31. Sinner, M.; Puls, J. J Chromatogr 1978, 156, 197.
32. http://www.sciencorp.com/pages/scion_image_windows.htm. Accessed date: March 14, 2009.
33. Evans, R. Appita J 1999, 52, 283.
34. Katuscak, S.; Kucera, L. Wood Res 2000, 45, 9.
35. Hrcka, R. Wood Res 2008, 53, 119.
36. Raya, S. S.; Yamada, K.; Okamoto, M.; Ueda, K. Polymer 2003, 44, 857.
37. Liese, W.; Bauch, J. Holz Roh Werkst 1977, 35, 267.
38. Hansmann, C.; Gindl, W.; Wimmer, R.; Teischinger, A. Wood Res 2002, 47, 1.
39. Yeh, T. F.; Braun, J. L.; Goldfarb, B.; Chang, H.-M.; Kadla, J.-F. Holzforschung 2006, 60, 1.
40. Raj, R. G.; Kokta, B. V.; Maldas, D.; Daneault, C. J Appl Polym Sci 1989, 37, 1089.
41. Herrera, P. J.; Valadez-Gonzalez, A. In Natural Fibers, Biopolymers, and Biocomposites; Mohanty, A., Misra, M., Drzal, L. T., Eds.; Taylor&Francis Group: 2005; p 189.
42. Mathew, A. P.; Oksman, K.; Sain, M. J Appl Polym Sci 2006, 101, 300.
43. Kokkonen, P.; Fardim, P.; Holmbom, B. Nordic Pulp Paper Res 2004, 19, 318.
44. Bengtsson, M.; Oksman, K. Compos Sci Technol 2006, 66, 2177.
45. Chartoff, R. P.; Weissman, P. T.; Sircar, A. In Assignment of the Glass Transition; Seyler, R. J., Ed.; ASTM International: Atlanta, 1994; p 88.
46. Pilla, S.; Gong, S.; O'neill, E.; Rowell, R. M.; Krzysik, A. M. Polym Eng Sci 2008, 48, 578.
47. Huda, M. S.; Drzal, L. T.; Mohanty, A. K.; Misra, M. Comp Sci Technol 2008, 68, 424.
48. Wang, H.; Sun, X.; Seib, P. J Appl Polym Sci 2001, 82, 1761.
49. Zhong, W.; Ge, J.; Gu, Z.; Li, W.; Chen, X.; Zang, Y.; Yang, Y. J Appl Polym Sci 1999, 74, 2546..

Publication C

Hrabalova, M., Schwanninger, M., Wimmer, R., Gregorova, A., Zimmermann, T. & Mundigler, N.

Fibrillation of Flax and Wheat Straw Cellulose and Its Effect on Thermal, Morphological and Viscoelastic Properties of Poly(vinylalcohol)/Fibre Composites.

BioResources, 2011, 6, 2, 1631-1647.

FIBRILLATION OF FLAX AND WHEAT STRAW CELLULOSE: EFFECTS ON THERMAL, MORPHOLOGICAL, AND VISCOELASTIC PROPERTIES OF POLY(VINYLALCOHOL)/FIBRE COMPOSITES

Marta Hrabalova,[a,]* Manfred Schwanninger,[b] Rupert Wimmer,[a,c] Adriana Gregorova,[d] Tanja Zimmermann,[e] and Norbert Mundigler [a]

Nano-fibrillated cellulose was produced from flax and wheat straw cellulose pulps by high pressure disintegration. The reinforcing potential of both disintegrated nano-celluloses in a polyvinyl-alcohol matrix was evaluated. Disintegration of wheat straw was significantly more time and energy consuming. Disintegration did not lead to distinct changes in the degree of polymerization; however, the fibre diameter reduction was more than a hundredfold, creating a nano-fibrillated cellulose network, as shown through field-emission-scanning electron microscopy. Composite films were prepared from polyvinyl alcohol and filled with nano-fibrillated celluloses up to 40% mass fractions. Nano-fibrillated flax showed better dispersion in the polyvinyl alcohol matrix, compared to nano-fibrillated wheat straw. Dynamic mechanical analysis of composites revealed that the glass transition and rubbery region increased more strongly with included flax nano-fibrils. Intermolecular interactions between cellulose fibrils and polyvinyl alcohol matrix were shown through differential scanning calorimetry and attenuated total reflection-Fourier transform infrared spectroscopy. The selection of appropriate raw cellulose material for high pressure disintegration was an indispensable factor for the processing of nano-fibrillated cellulose, which is essential for the functional optimization of products.

Keywords: Polyvinyl alcohol; Nano-fibrillated cellulose; Composite film; Viscoelastic properties; Thermal properties; Film; ATR-FTIR

*Contact information: a: Institute for Natural Materials Technology, IFA-Tulln, University of Natural Resources and Life Sciences, Vienna, Konrad Lorenz Strasse 20, A-3430 Tulln, Austria; b: Department of Chemistry, BOKU – University of Natural Resources and Life Sciences, Vienna, Muthgasse 18, A-1190 Vienna, Austria; c: Faculty of Forest Sciences and Forest Ecology, University of Göttingen, Büsgenweg 4, 37077 Göttingen, Germany; d: Institute for Chemistry and Technology of Materials, Graz University of Technology, Stremayrgasse 9, 8010 Graz, Austria; e: Empa, Swiss Federal Laboratories for Materials Science and Technology, Wood Laboratory, Ueberlandstrasse 129, CH-8600 Dübendorf, Switzerland; *Corresponding author: martahrabalova@hotmail.com*

INTRODUCTION

Cellulose is the most abundant naturally occurring polymer on earth. The average plant contains roughly 40% of cellulose, which adds up to bio-based resource of approximately 2000 million dry tons (Rowell et al. 2000). During the past years a lot of research on the disintegration of cellulose fibres through chemical (Samir et al. 2005) or mechanical treatments such as high pressure disintegration (Siró and Plackett 2010) has been performed. Due to the absence of imperfections that originated from chain folding, the obtained fibres should demonstrate unique properties similar to the perfect crystal of native cellulose (Nakagaito & Yano 2005). Chemical treatments have led to uniformly sized cellulose whiskers of lower aspect ratios (Favier et al. 1995; Helbert et al. 1996), while mechanical refining or homogenization processes results in nano-fibrillated cellulose (NFC) fibrils with a wide size distribution and also higher aspect ratio (Nakagaito and Yano 2004). Today, high pressure fibrillation is the most common method for producing nano-sized cellulose materials with applications in food industry, cosmetics or pharmaceutical industry (Nakagaito and Yano 2005). Structural features of cellulosic fibres and the ability to form a variety of intermolecular interactions with other materials are fundamental aspects for good adhesion between fibre and matrix, which greatly improve performance of fibre reinforced composites (Iwamoto et al. 2007; Lu et al. 2008; Schartel et al. 1996).

With the use of hydrophilic-type matrices such as polyvinyl alcohol (PVA) higher compatibility as well as better fibre dispersion is possible. In addition, the obtained composite material is environmentally friendly and processable at ambient conditions (Yu et al. 2006). Due to their hydrophobicity PVA/nano-cellulose composites are suitable for the production of medical device applications.

PVA is a water-soluble and highly biodegradable polymer prepared through hydrolysis of polyvinyl acetate. PVA has excellent antistatic and oxygen barrier properties and it is resistant to most organic solvents (Moore et al. 1997). Due to its hydrophilicity PVA is also suitable for the production of nano-composites that are destined for biomedical applications (Millon & Wan 2006; Galya, et al. 2008). NFCs made from different raw materials have been utilized with PVA as a matrix, including commercially microfibrillated cellulose (Lu et al. 2008), nano-fibrillated wood (Zimmermann et al. 2004), nano-fibrillated sugar beet pulp (Leitner et al. 2007) or fibrils from Swede roots (Bruce et al. 2005) NFCs. It was shown that the choice of fibre raw material affects processing parameters, including energy consumption, the latter being critical when used industrially (Zimmermann et al. 2010). The objectives of this work are (1) to apply the fibrillation process to two types of cellulose fibres, (2) to produce nano-composites with the two reinforcing fibre types in PVA matrices, (3) to determine thermal, morphological and viscoelastic properties of obtained composite materials, and (4) to determine water uptake sensitivity at a given climate change.

It is known that different types of cellulose may show a diversity of functions in composites (Klemm et al. 2009). Therefore, cellulose pulps from flax (F) and wheat (WS) were used as raw materials. The two nano-fibrillated fibre types were incorporated into PVA matrices at NFC mass fractions (w_C) ranging between 5 and 40%. The impact of a wide range of NFC content on properties of PVA/NFC composites was assessed. Reinforcing potential of NFCs for hydrophilic PVA matrix was particularly determined by dynamic mechanical analysis (DMA). Further thermal properties of PVA/NFC were observed using differential scanning calorimetry (DSC). Cellulose dispersion in the PVA matrix was studied through scanning electron microscopy (NanoSEM); and intermolecular H-bondings were identified through Attenuated Total Reflection-Fourier Transform InfraRed Spectroscopy (ATR-FTIR). Cellulose polymerization degree (DP) was determined by viscosimetry, and structural order was assessed by ATR–FTIR.

Cellulose structures were identified through field emission scanning electron microscopy (FE-SEM).

EXPERIMENTAL

Materials

Poly(vinyl alcohol) Airvol 523 (medium viscosity, degree of hydrolysis 86% to 88% and average molecular weight ~ 89,000) was supplied by AIR Products and Chemicals, Inc. (USA). For the isolation of nanofibrillated cellulose two commercially available pulps obtained from wheat straw (*Tritium vulgare*) and flax (*Linum usitatissiumum*) were provided by the Delfort group AG (Austria).

Methods

High pressure disintegration of cellulose fibres

Both pulp types were swollen in deionised water in a 10 L thermostatic reactor and stirred. To obtain a more homogeneous suspension, a big-sized Ultra-Turrax (T 50 BASIC IKA WERKE) was used. The suspension was then passed through an inline dispersion system, equipped with the Ultra-Turrax Megatron MT 3000 (Kinematica AG, Switzerland), which was attached to a thermostatic reactor. Inline dispersion was performed between 15,000 rpm and 20,000 rpm to disintegrate the fibres. The temperature of the system was kept at 15 °C to avoid growth of bioactive organisms. Finally, the fibre suspension was passed through a high-pressure homogenizer (Microfluidizer, M-110Y). The exact apparatus configuration is described elsewhere (Zimmermann et al. 2010). The resulting suspensions were centrifuged several times at 5,000 rpm for 20 minutes at temperature of 20 °C, until 5.95% dry matter content was reached for the nanofibrillated wheat straw pulp (WS-NFC) and 9.75% for nanofibrillated flax pulp (F-NFC).

Viscosimetric measurements

The intrinsic viscosity of both non-fibrillated and fibrillated celluloses was determined according to ISO 5351 standard (Anonymous 2004). Degree of polymerization (DP) was then calculated using the Staudinger-Mark-Houwink equation,

$$[\eta] = (K*DP)^a \tag{1}$$

where the constants K and a depend on the polymer-solvent system (Henriksson 2008). For wheat straw $K = 2.28$ and $a = 0.76$, while for flax $K = 0.42$ and $a = 1$, respectively (Marx-Figini 1978).

Field emission scanning electron microscopy (FE-SEM)

Aqueous cellulose suspensions of 0.1% cellulose mass fraction were dropped on a specimen holder and directly sputtered with a 7.5 nm thick platinum layer (BAL-TEC MED 020 modular high vacuum coating systems, BALTEC AG, Principality of Liechtenstein). Microscopic analysis was accomplished for all cellulose materials prior and post disintegration using a Jeol 6300F (Jeol Ltd., Japan), operated under high vacuum at 5kV beam voltage. Diameter of the fibres was determined from SEM images with 20,000 and 10,000 magnifications, respectively. Images were analysed using Adobe Acrobat 7.0 Professional software, and averages were built from measurement of 200 fibres.

Composite preparation

A 3.3%-PVA solution was prepared by stirring PVA in deionised water at room temperature for 6 hours. The suspension with w_C = 0.5% of NFC was prepared with deionized water by stirring the cellulose for 6 hours. The suspension was treated by a touch mixer (VORTEX-GENIE 1 Touch Mixer) for 5 minutes and dispersed with the Ultra-Turrax IKA T10 basic for 2 minutes to achieve high homogeneity. PVA solution was mixed with appropriate amounts of NFC aqueous suspension to prepare composite films with a final NFC contents ranging between 0 and 40%. The PVA/NFC suspensions were stirred for 16 hours, then cast in Teflon forms and finally dried at 50% relative humidity (RH) and 23°C up to 7 days. In addition to PVA/NFC films neat cellulose films were also cast from a w_C = 0.5% non-fibrillated cellulose suspension, as well as from a 0.5% NFC suspension. After water evaporation, all films were dried under vacuum for 24 hours at 40 °C. Prior to testing, the obtained composite sheets were kept in a desiccator with dried silica gel to avoid water swelling and consequently plastification of the PVA matrix. The different sample groups were marked as follows: F-NFC for neat nano-fibrillated flax cellulose films, WS-NFC for neat nano-fibrillated wheat straw cellulose films, and PVA for neat polyvinyl alcohol film. PVA/xF-NFC, PVA/xWS-NFC stand for PVA composite films with nano-fibrillated flax and wheat straw cellulose, respectively, having a x-percent mass fraction of NFC in the composite.

ATR-FTIR spectroscopy

The ATR-FTIR spectra of films were measured with 32 scans per sample at a spectral resolution of 4 cm^{-1} and a wavenumber range between 4000 cm^{-1} to 600 cm^{-1}. An attenuated total reflectance (ATR) device (MIRacle™, from Pike Technologies, www.piketech.com) was installed on a Bruker Vertex 70 device (Bruker Optics, www.brukeroptics.de), together with a DLATGS mid-infrared detector. Four spectra were recorded for each film and averaged. For the calculation of relative crystallinity the spectra obtained from cellulose films were minimum-maximum normalized over the whole wavenumber range. Baseline correction was performed by the rubber-band method to minimize differences between spectra due to baseline shifts (Erukhimovitch et al. 2006). Composition dependent spectra of O-H stretching region were minimum-maximum corrected in the wavenumber range between 3800 cm^{-1} and 3000 cm^{-1}.

Scanning electron microscopy (NanoSEM)

Composite films were cryo-fractured in liquid nitrogen and fracture surfaces were sputtered with 7.5 nm platinum layer. Fractured surfaces of the PVA/NFC composites were studied with a reflection electron microscope, FEI Nova NanoSEM 230.

Water-uptake
Relative water uptake measurements were performed on films sized 20 mm x 20 mm and a thickness of about 0.20 mm. Dry weight of each sample was over 100 mg. After dry weighing, the samples were placed in a conditioned room with 50% relative humidity and 23°C. Weight changes of the nano-fibrillated celluloses, PVA and the PVA/NFC composites samples were measured up to 9 days with the moisture content determined as follows,

$$wt.\% = 100 * \frac{W_2 - W_1}{W_1} \qquad (2)$$

where W_1 is the weight of the dry sample, and W_2 the weight of the wet sample.

Differential scanning calorimetry (DSC)
Melting behaviour of the PVA and PVA/NFC composite films was studied on a NETZSCH DSC 200 F3 Maja. Approximately 8 mg of sample was sealed in pierced aluminium pans. Measurements were conducted within the temperature range of 20 °C and 220 °C at a heating rate of 10 K/min, and a nitrogen gas flow of 20 mL/min. Melting temperature (T_m) was derived from the melting peak. Heat of fusion (ΔH_m) was evaluated from the peak area with respect to the PVA mass fraction in the composite film. Results are reported as average values of duplicate measurements. The maximal experimental error of T_m data was indicated to be ±0.5°C and ±1 J/g of ΔH_m.

Dynamic-mechanical analysis (DMA)
Viscoelastic properties of PVA/NFC composite films were measured on a NETZSCH 242 C in tensile mode. Strips of 10 mm × 6 mm × 0.1 mm in size were cut. Storage modulus (E'), and the height of loss tangent peak (tan δ) were measured within the temperature range of -20 °C to 120 °C at a strain sweep frequency of 1 Hz and a heating rate of 3 °C/min. Glass transition temperature (T_g) was evaluated from the E' inflection point.

RESULTS AND DISCUSSION

Processibility of nano-fibrillated cellulose
Flax cellulose was chosen because of its excellent mechanical properties and outstanding reinforcement potential (Charlet et al. 2010), while wheat straw cellulose as residual agricultural material seemed to be an economically and environmentally viable source of cellulose (Liu et al. 2005). Processing conditions for each material are listed in Tables 1 and 2.

Cellulose pre-processing involved the swelling followed by several disintegration steps. Pre-processing procedures were necessary to obtain homogeneous cellulose suspensions, making it possible to pass through the high pressure homogenizer. As water penetrated the amorphous cellulose regions, hydrogen bonds were disrupted and distances between cellulose chains loosened (Tímár-Balázsy and Eastop 2002), which eased mechanical disintegration. The compliancy of the wheat straw slurry to pass the inline disperser turned out to be low; therefore, the time of WS swelling was 4 times longer than for flax. In addition, due to repeated chamber clogging when passing WS through the high pressure homogenizer, the required time for inline dispersion of the WS slurry was also markedly longer than for the F slurry. To lower the chance of chamber clogging the WS slurry was diluted to a mass fraction of 0.375%.

Table 1. Processing Conditions upon Mechanical Pre-treatment

Material	Swelling		Ultra-Turrax (big)		Inline dispersion	
	Cellulose mass fraction in slurry [%]	Time [h]	Cellulose mass fraction in slurry [%]	Processing time [h]	Cellulose mass fraction in slurry [%]	Processing time [h]
F	1	62	1	5	1	6.5
WS	0.8	254	0.8	6	0.8	9.2

Table 2. Chamber Combinations Used during High Pressure Disintegration.

Chamber combinations	F slurry (No. of passes)	WS slurry (No. of passes)
400 μm + 200 μm	12	14
400 μm + 75 μm	0	8
200 μm + 75 μm	12	0

The mass fraction of F and WS in slurry during processing was 1% and 0.375%, respectively.

During the final high pressure disintegration step (using the interaction chamber combinations 200 μm + 75 μm for F, and 400 μm + 75 μm for WS), 100 g of flax dry matter has passed through within 18 minutes, whereas the same amount of WS took 70 minutes. For both celluloses the high pressure fibrillation was interrupted on the basis of visual evaluation as soon the suspension was homogeneous and free of agglomerates. The absence of agglomerates was continuously checked by rubbing the suspension between two fingers (Zimmermann et al. 2010).

Table 3. Intrinsic Viscosity $[\eta]$ and DP of Wheat Straw and Flax Celluloses

Material	$[\eta]$ [mL/g]	DP
F	300	714
F-NFC	290	691
WS	710	1,909
WS-NFC	640	1,665

Degree of polymerization (DP)

Through viscosimetric measurements the DPs for both raw materials were obtained. Overall, the DP of flax cellulose was markedly lower than that of wheat straw cellulose (Table 3). As DP predicates, the stiffness of cellulose chain having higher DP for WS cellulose might explain the difficulties present during processing. Changes of DP induced by fibrillation were rather small for both materials (3.2% for F and 12.8% for WS) and indicate breakage along the fibre (Iwamoto et al. 2007). The DP drop due to high pressure disintegration was more pronounced for higher DP material, which was in accordance with the findings of Zimmermann et al. (2010).

Morphology of the fibres

The structural nature of the non-fibrillated F and WS celluloses and effects of high pressure treatment on their structure are shown in Fig. 1. Fibre diameter prior to fibrillation was around 12 μm for F cellulose, and 18 μm for WS cellulose. After high pressure disintegration the obtained fibrils had diameters markedly different from the previous fibres. For F-NFC the fibril diameters ranged between 37 nm and 430 nm (average 81 nm), and for WS-NFC between 60 nm and 1,080 nm (average 128 nm). Fraction of fibrils reaching nanometer dimension (<100nm) was determined to be 82% for F-NFC, and to be 58% for WS-NFC, respectively. The increased variability was due to the fact that the fibrils remained partly as agglomerates. Even with the presence of agglomerated fibrils the homogenization process greatly increased the surface area.

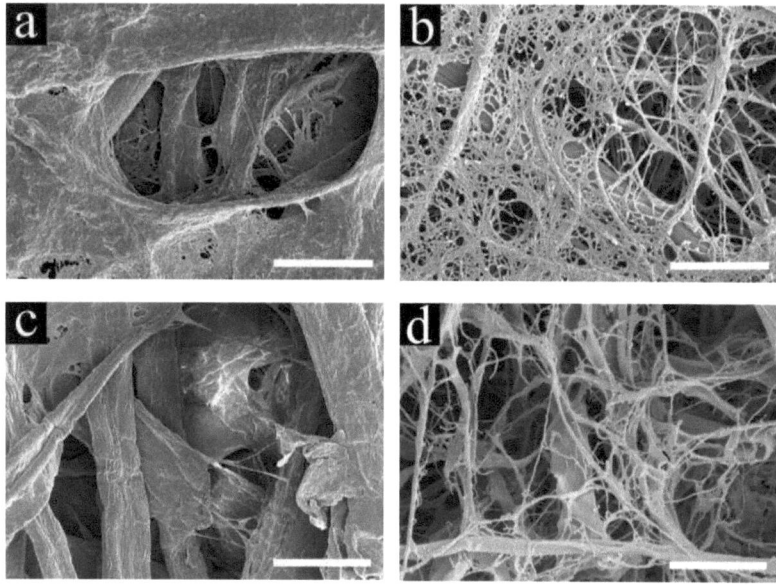

Fig. 1. FE-SEM micrographs of a) F cellulose non-fibrillated, b) F-NFC, c) WS cellulose non-fibrillated, d) WS-NFC. The white scale bar represents the length of 20 μm (a and c) and 3 μm (b and d)

Table 4. Changes of Relative Crystallinity due to Fibrillation Determined by ATR-FTIR

Cellulose	$A_{1400-1289}/A_{898}$
F	5.8
F-NFC	5.4
WS	4.0
WS-NFC	3.8

ATR-FTIR analysis of celluloses and composites

Normalized spectra of NFC and non-fibrillated cellulose sheets acquired for the wavenumber range 1800 cm^{-1} to 600 cm^{-1} are shown in Fig. 2. Spectral profiles in this region turned out to be very similar; however, band intensities were slightly changed. Relative crystallinity of both materials before and after high pressure disintegration was calculated from band area ratios $A_{1400-1289}/A_{898}$ (Schwanninger et al. 2004). Schultz et al. (1985) report that this ratio correlates well with the crystallinity determined by X-ray diffraction. As can be seen in Table 4, relative crystallinity was found to be higher for the flax cellulose films. Crystallinity decreased after high pressure disintegration for both materials. Similar findings were reported by Iwamoto *et al.* (2007) who measured decline of cellulose crystallinity through X-ray diffraction as a response of high pressure disintegration. Schwanninger *et al.* (2004) reported a decrease in cellulose relative crystallinity due to the ball milling measured by FTIR, which is another confirmation for their own findings.

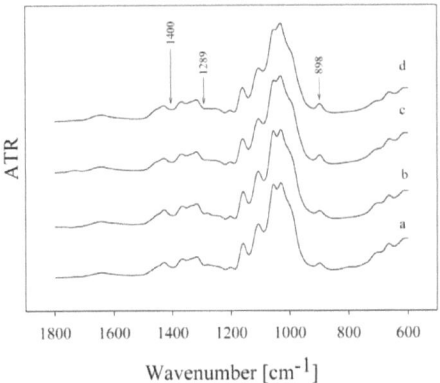

Fig. 2. ATR-FTIR spectra in the range from 1800 cm^{-1} to 600 cm^{-1} of NFC and non-fibrillated cellulose films. a) F-NFC, b) F cellulose non-fibrillated, c) WS-NFC and d) WS cellulose non-fibrillated. (Absorbance scale is not given because spectra are shifted parallel.)

FT-IR is a suitable method for the determination of molecular interactions among various chemical groups and consequently the composite compounds (Kondo et al. 1994; Sawatari and Kondo 1999; Shibayama et al. 1991; Zhu et al. 2003). In PVA/cellulose systems mutual interactions via hydrogen bonding of hydroxyl groups have been reported (Kondo et al. 1994; Shibayama et al. 1991). Using FTIR hydrogen bonding between OH groups can be observed within the O-H stretching range of 4000 cm^{-1} to 3000 cm^{-1}. Normalized FTIR spectra of the O-H stretching region for NFC, PVA, and PVA/NFC with up to $w_C = 40\%$ are shown in Fig. 3, for both nano-composite types. The maximum of cellulose O-H stretching band was situated at 3336 cm^{-1} (at 3342 cm^{-1} O(3)-H(3)···O(5) intra-chain H-bond (Maréchal and Chanzy 2000)), which is close to the hydroxyl stretching vibration of PVA (3325 cm^{-1}) (Tadoroko 1959). The shift of the O-H stretching peak to the lower wavenumber of 3315 cm^{-1} was observed for PVA/NFC systems containing 5% of cellulose. This shift was accompanied by a broadening of the O-H stretching band in the case of the WS composite. The extended width of O-H stretching band refers to a broader distribution of hydrogen bond lengths, and the shift to a lower wavenumber was due to bond strengthening and shortening. The shift of O-H stretching bands of the composite to a lower wavenumber relative the NFCs and PVA peaks was expected for the composite having a low NFC mass fraction.

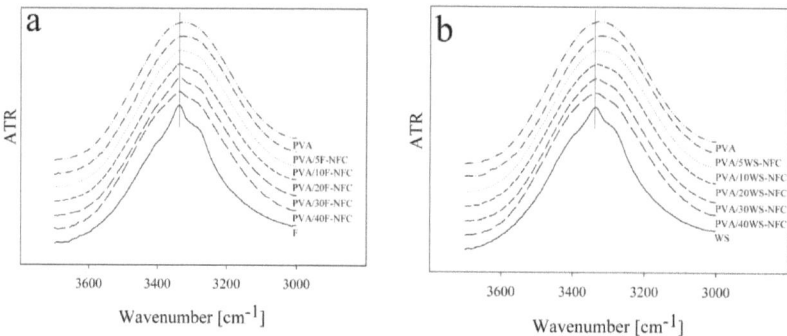

Fig. 3. Composition-dependent ATR-FTIR spectra of O-H stretching region of a) PVA, F-NFC, and PVA/F-NFC, and b) PVA, WS-NFC, and PVA/WS-NFC composites. (Absorbance scale is not given because spectra are shifted parallel.)

With increasing NFC mass fractions the O-H stretching bands became narrower and gradually shifted to higher wavenumbers, up to a wavenumber that corresponded to the cellulose O-H stretching band (3336 cm^{-1}). This peak position was reached at $w_C = 20\%$ F-NFC, and at $w_C = 30\%$ of WS-NFC, respectively. Narrowing of O-H stretching bands indicated a more regular structure of the composite with even distances of H-bonds or a reduction of interactions between PVA and NFC. This was probably covered by interactions within the neat materials, combined with a decline in PVA crystallinity. With increasing NFC content in the PVA matrix the shoulder assigned to O(6)-H(6)···O(3) intermolecular H-bonding in the cellulose NFC (3270 cm^{-1}) (Fengel 1993) became more evident.

Morphology of the composites

Scanning electron microscopy of cryo-fractured surfaces of PVA/NFC composites was done to characterize cellulose dispersion in the PVA matrix (Fig. 4). No voids signifying cellulose pull-outs were observed for composites with a cellulose content up to $w_C = 20\%$ (Fig. 4 a-d). The fracture surfaces appeared smooth. This was not observed for composites having higher cellulose contents (Fig. 4 e, f). It is also seen from the NFC fibre pullouts in PVA matrix that WS-NFC fibres were bigger than F-NFC fibres. The latter were homogeneously dispersed in the PVA matrix. WS-NFC in PVA matrix tended to agglomerate within the entire mass fraction range and its dispersion was poorer. The arrow in Fig. 4 d marks the region of WS-NFC aggregation/sedimentation in the PVA matrix. Dispersion of cellulose in aqueous suspension should improve along with the degree of cellulose refinement (Seydibeyoğlu and Oksman 2008). Therefore, the poorer dispersion of WS-NFC in PVA compared to PVA/F-NFC composite is most likely related to a coarser structure of WS-NFC.

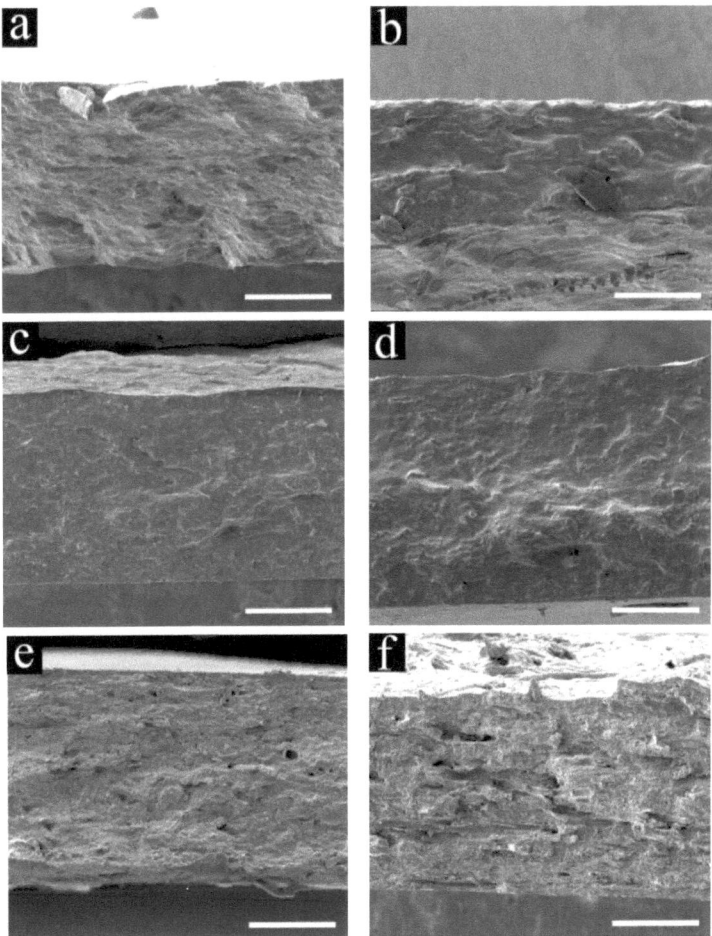

Fig. 4. SEM micrographs of cryogenic fracture surfaces PVA/NFC films a) PVA/10F-NFC, b) PVA/10WS-NFC, c) PVA/20F-NFC, d) PVA/20WS-NFC, e) PVA/40F-NFC, f) PVA/40WS-NFC. The white scale bar represents the length of 50 μm

Table 5. Results of Relative Water Uptake at Equilibrium and the Time Needed to Reach Equilibrium Swelling

Sample	Time [h]	wt. %	Sample	Time [h]	wt. %
PVA	48	5.2	PVA	48	5.2
PVA/5F-NFC	6	7.4	PVA/5WS-NFC	6	7.1
PVA/10F-NFC	48	6.2	PVA/10WS-NFC	48	6.3
PVA/20F-NFC	48	5.8	PVA/20WS-NFC	48	5.7
PVA/30F-NFC	48	5.4	PVA/30WS-NFC	48	5.4
PVA/40F-NFC	48	4.9	PVA/40WS-NFC	48	5.1
F-NFC	30	3.6	WS-NFC	30	4.3

Water sensitivity

Table 5 gives the results for relative water uptake at 50% relative humidity and 23°C, with the time needed to reach equilibrium. Water uptake was performed to observe the water sensitivity of the films at moderate humidity (50%).

Data revealed that PVA films were more sensitive to moisture than NFC films, with no marked differences between F-NFC and WS-NFC. PVA/NFC films having 5% NFC mass fraction were most sensitive to moisture, and equilibrium moisture content was reached already after 6 hours of exposure. For both NFC types the relative water uptake tended to decline as the NFC mass fraction went up. The decrease of water uptake might be linked to the formation of a continuous cellulosic structure in PVA matrix and thus a less disturbed and less moisture accessible PVA structure by contrast to the films with 5% NFC mass fraction. Similar results were reported for starch/tunicin whiskers nanocomposites. With an increasing content of whiskers in starch matrix the water uptake at equilibrium decreased, which was linked to a higher formation of a microfibrillar network (Anglès and Dufresne 2000). The formation of cellulose continuous structure was also proposed on the basis of ATR-FTIR measurements by observing the OH stretching region. The effect of different RH on properties of PVA/nanocellulose composites was described elsewhere (Roohani et al. 2008).

Thermal properties of the composites

DSC measurements were accomplished for the PVA and PVA/NFC films. Figure 5 shows that changes in melting temperatures (T_m) of the composite films as well as heat of fusion (ΔH_m) were similar across the different compositions, irrespective of the used NFC reinforcement. Heat of fusion was calculated with respect to the PVA content in the composite. For composites having w_C = 5% and 10%, respectively, ΔH_m increased compared to the neat PVA. This can be ascribed to the nucleating ability of the fibrils (Lu et al. 2008). With higher amounts of NFCs ΔH_m of PVA matrix unambiguously declined. As can be seen in Figure 5, the melting temperature varied only slightly. T_m reflects the nature of the crystalline structure. Lowering of T_m is a consequence of restricted spherulite growth or crystalline imperfection. However, in some composite systems, especially in good miscible systems, the decrease of melting temperature confirms specific interactions between the blended components (Dubief et al. 1999; Nishio et al.

1989; Roohani et al. 2008; Samir et al. 2004). T_m decreased despite the increasing ΔH_m. Therefore, due to the similar nature of filler and matrix, it is reasonable to ascribe the drop of T_m to interactions between composite components. This result agreed well with the interaction revealed by ATR-FTIR.

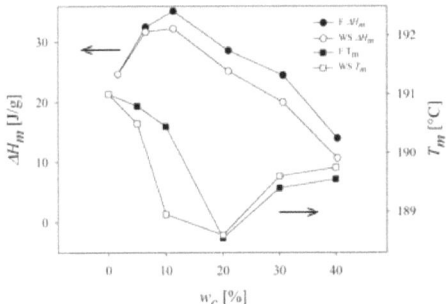

Fig. 5. Melting temperatures (T_m) and heat of fusion (ΔH_m) of the PVA matrix in relation to the cellulose mass fraction

Viscoelastic properties of the composites

The study of viscoelastic properties was focused on the relaxation range close to the glass transition of PVA. As shown in Fig. 6, the concentration-dependant effect of NFC type on viscoelastic properties of PVA/NFC composites is obvious in every single region of viscoelastic behaviour. Table 5 provides numerical results for dynamic-mechanical analyses, i.e. storage modulus at 20 °C and 100 °C, respectively. The table also lists T_g as well as tan δ height for both PVA and PVA/NFC films. In the glassy region even the presence of 5% mass fraction of NFC resulted in remarkably higher mechanical properties of the composite, compared to neat matrix. E' at 20 °C was 1.7 and 1.8 times higher for the F-NFC and WS-NFC reinforced PVA composites, respectively, compared to the PVA reference (Table 6). Upward drift of the E' in the glassy region with increasing mass fraction of NFC was observed for both materials. With a reinforcement between w_C = 30% and 40% the WS-NFC reinforcement was distinctly stronger than F-NFC, and reached almost a threefold improvement over neat PVA. Higher E' in the glassy region might be a consequence of higher DP of WS-NFC, which is a known factor for cellulose stiffness.

The stiffness increase in the rubbery region (E' at 100 °C) was distinct for both materials and reached more than a thirtyfold increased with w_C = 10%, exceeding the initial value of PVA by over a factor of 100 with w_C = 40%. The mechanical restraint in the rubbery region was more pronounced for the PVA/F-NFC compared to PVA/WS-NFC. Such behaviour was described by others and the improvement of mechanical performance as well as the thermal stability was attributed either to the percolation effect, or to a tangling effect of cellulose microfibrils (Dubief et al. 1999; Samir et al. 2004). Another explanation is the existence of intermolecular interactions between cellulose and the polymer matrix (Shibayama et al. 1991). However, better performance of PVA/F-

NFC composites in the rubbery region is most likely based on better dispersion of the F-NFC in the PVA matrix, as demonstrated through SEM (cp. Fig. 4).

Table 6. DMA Results: Storage Modulus in Glassy and Rubbery Regions, Glass Transition and Height ff Loss Tangent Peak of PVA, PVA/F-NFC, and PVA/WS-NFC Composites

Sample	E' at 20 ℃ [GPa]	E' at 100 ℃ [GPa]	T_g [℃]	tan δ (height)
PVA	2.42	0.01	50.5	0.456
PVA/5F-NFC	4.12	0.21	51.7	0.335
PVA/10F-NFC	4.39	0.46	56.4	0.291
PVA/20F-NFC	5.84	0.90	58.8	0.205
PVA/30F-NFC	5.42	1.12	60.7	0.203
PVA/40F-NFC	6.18	1.92	60.9	0.114
PVA/5WS-NFC	4.41	0.17	53.9	0.359
PVA/10WS-NFC	4.81	0.39	53.8	0.261
PVA/20WS-NFC	5.59	0.59	58.4	0.296
PVA/30WS-NFC	6.66	0.80	58.4	0.193
PVA/40WS-NFC	7.16	1.51	58.3	0.187

As shown in Fig. 6 the relaxation region was broadened with shifts to higher temperatures as NFC mass fraction increased. The effect of crystallinity on E' was blocked due to decreasing PVA crystallinity with NFC content. T_g of PVA/WS-NFC composites leveled off at 58 °C and w_C = 20%; and T_g of PVA/F-NFC composite equalized at w_C = 30%, reaching almost 61 °C, which was 10 °C above that of neat PVA. The strong decrease of the tan δ peak as a consequence of cellulose insertion into the PVA matrix is an indicator of good adhesion between filler and matrix (Gregorova et al. 2010; Hrabalova et al. 2010; Keusch and Haessler 1999; Pothan et al. 2003). Interestingly, the height of the tan δ peak was similar irrespective to the used NFC type. This result might lead to the conclusion that hydrogen bonding formed within the composite is a decisive factor impacting tan δ. The effect of hydrogen bonding on tan δ was reported earlier by, e.g., Huda et al. (2008).

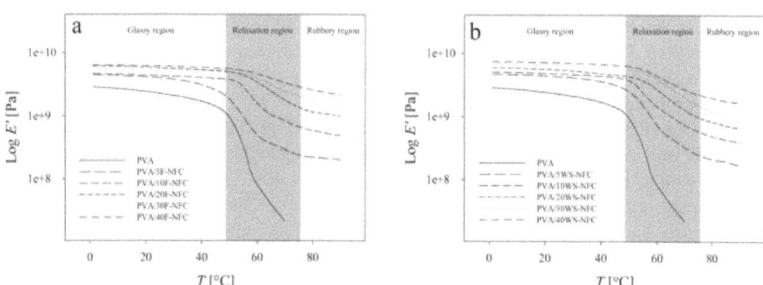

Fig. 6. Logarithm of the storage modulus E' versus temperature at 1Hz for a) PVA/F-NFC and b) PVA/WS-NFC at NFC mass fractions in the range between 5 and 40%

CONCLUSIONS

Disintegration of flax and wheat straw cellulose pulps by a high pressure homogenizator was investigated. Materials were prepared in order to evaluate the effect of nanofibrillated celluloses (NFCs) on the properties of their polyvinyl alcohol (PVA) composites. The suitability of flax cellulose pulp and wheat straw cellulose pulp for high pressure disintegration was investigated, as well as the reinforcing potential of disintegrated NFCs.

1. It was shown that the production of NFC from wheat straw (WS-NFC) might be inferior to that obtained from flax without additional pre-treatment of the raw material (e.g. chemical). Such production is technically demanding, tedious, and time-consuming. The less demanding disintegration of flax cellulose was most probably due to the lower DP of its cellulose component.
2. The PVA/WS-NFC showed a better storage modulus in glassy region only. The better mechanical performance of the flax-containing PVA/F-NFC composites in rubbery region was most probably related to better dispersion of F-NFC in the PVA matrix. T_g, determined from DMA, was improved due to the presence of NFC. T_g increased with w_c for both materials.
3. As concluded from SEM micrographs, the more homogeneous dispersion in PVA was better reached with F-NFC, as compared to the WS-NFC dispersion. The better dispersion of F-NFC is explained by its higher fineness.
4. Water sensitivity measurements showed that at 50% RH and 23°C the most sensitive composites were those having 5% NFC. Water sensitivity declined at higher NFC mass fractions.
5. Inter-polymer bonding between NFCs and PVA hydroxyl groups was supported by DSC and ATR-FTIR. Both methods showed inter-polymer bonding as most pronounced up to $w_C = 20\%$ NFC. It is suggested that at higher NFC contents ($w_C > 20\%$) interactions within each component apart are more pronounced than inter-polymer interactions between PVA and NFCs.

ACKNOWLEDGEMENTS

This work was financed by Austrian Science Fund (FWF, Project L319-B16) and COST Action E50 CEMARE (ref. no. COST-STSM-E50-4325). Thanks are due to Esther Strub from the Wood Laboratory at EMPA, Dübendorf, Switzerland, for her help with viscosity measurements and FE-SEM.

REFERENCES CITED

Anglès, M. N., and Dufresne, A. (2000). "Plasticized starch/tunicin whickers nanocomposites. 1. Structural analysis," *Macromolecules* 22(33), 8344-8353.

Anonymous (2004). "Pulps – Determination of limiting viscosity number in cupriethylenediamine (CED) solution," In ISO. *ISO 5351*: ISO.

Bruce, D. M., Hobson, R. N., Farrent, J. W., and Hepworth, D. G. (2005). "High-performance composites from low-cost plant primary cell walls," *Composites Part A: Applied Science and Manufacturing* 36(11), 1486-1493.

Charlet, K., Jernot, J. P., Eve, S., Gomina, M., and Bréard, J. (2010). "Multi-scale morphological characterisation of flax: From the stem to the fibrils," *Carbohydrate Polymers* 82(1), 54-61.

Dubief, D., Samain, E., and Dufresne, A. (1999). "Polysaccharide microcrystals reinforced amorphous poly(beta-hydroxyoctanoate) nanocomposite materials," *Macromolecules* 32(18), 5765-5771.

Erukhimovitch, V., Talyshinsky, M., Souprun, Y., and Huleihel, M. (2006). "FTIR spectroscopy examination of leukemia patients plasma," *Vibrational Spectroscopy* 40, 40-46.

Favier, V., Chanzy, H., and Cavaille, J. Y. (1995). "Polymer nanocomposites reinforced by cellulose whiskers," *Macromolecules* 28(18), 6365-6367.

Fengel, D. (1993). "Influence of water on the OH valency range in deconvoluted FTIR spectra of cellulose," *Holzforschung* 47, 103-108.

Galya, T., Sedlařík, V., Kuřitka, I., Novotný, R., Sedlaříková, T., and Sáha, P. (2008). "Antibacterial poly(vinyl alcohol) film containing silver nanoparticles: Preparation and characterization," *Journal of Applied Polymer Science* 110(5), 3178-3185.

Gregorova, A., Hrabalova, M., Kovalcik, R., and Wimmer, R. (2011). "Surface modification of spruce wood flour and effects on the dynamic fragility of PLA/wood composites," *Polymer Engineering & Science* 51, 143-150.

Helbert, W., Cavaille, J. Y., and Dufresne, A. (1996). "Thermoplastic nanocomposites filled with wheat straw cellulose whiskers. 1. Processing and mechanical behavior," *Polymer Composites* 17(4), 604-611.

Henriksson, M. (2008). "Cellulose nanofibril networks and composites. Preparation, structure and properties," *KTH Chemical Science and Engineering* (p. 61).

Hrabalova, M., Gregorova, A., Wimmer, R., Sedlarik, V., Machovsky, M., and Mundigler, N. (2010). "Effect of wood flour loading and thermal annealing on viscoelastic properties of poly(lactic acid) composite film," *Journal of Applied Polymer Science, in press* (DOI: 10.1002/app.32509).

Huda, M. S., Drzal, L. T., Mohanty, A. K., and Misra, M. (2008). "Effect of fiber surface treatments on the properties of laminated biocomposites from poly(lactic acid) (PLA) and kenaf fibers," *Compos Sci. Tech.* 68, 424.

Iwamoto, S., Nakagaito, A. N., and Yano, H. (2007). "Nano-fibrillation of pulp fibers for the processing of transparent nanocomposites," *Applied Physics A: Materials Science & Processing* 89(2), 461-466.

Keusch, S., and Haessler, R. (1999). "Influence of surface treatment of glass fibres on the dynamic mechanical properties of epoxy resin composites. *Composites Part A: Applied Science and Manufacturing* 30(8), 997-1002.

Klemm, D., Schumann, D., Kramer, F., Hessler, N., Koth, D., and Sultanova, B. (2009). "Nanocellulose materials – Different cellulose, different functionality," *Macromol. Symp.* 280, 60-71.

Kondo, T., Sawatari, C., Manley, R. S., and Gray, D. G. (1994). "Characterization of hydrogen-bonding in cellulose synthetic-polymer blend systems with regioselectively substituted methylcellulose," *Macromolecules* 27(1), 210-215.

Leitner, J., Hinterstoisser, B., Wastyn, M., Keckes, J., Gindl, W. (2007). "Sugar beet cellulose nanofibril-reinforced composites," *Cellulose* 14(5), 419-425.

Liu, R. G., Yu, H., and Huang, Y. (2005). "Structure and morphology of cellulose in wheat straw," *Cellulose* 12(1), 25-34.

Lu, J., Wang, T., and Drzal, L. T. (2008). "Preparation and properties of microfibrillated cellulose polyvinyl alcohol composite materials," *Composites Part A: Applied Science and Manufacturing* 39(5), 738-746.

Maréchal, Y., and Chanzy, H. (2000). "The hydrogen bond network in I_β cellulose as observed by infrared spectrometry," *Journal of Molecular Structure* 523, 183-196.

Marx-Figini, M. (1978). "Significance of the intrinsic viscosity ratio of unsubstituted and nitrated cellulose in different solvents," *Die Angewandte Makromolekulare Chemie*, 72, 161-171.

Millon, L. E., and Wan, W. K. (2006). "The polyvinyl alcohol-bacterial cellulose system as a new nanocomposite for biomedical applications," *Journal of Biomedical Materials Research Part B-Applied Biomaterials* 79B(2), 245-253.

Moore, G. F., and Saunders, S. M. (1997). *Advances in Biodegradable Polymers*. Rapra Publishing.

Nakagaito, A. N., and Yano, H. (2004). "The effect of morphological changes from pulp fiber towards nano-scale fibrillated cellulose on the mechanical properties of high-strength plant fiber based composites," *Applied Physics A: Materials Science & Processing* 78(4), 547-552.

Nakagaito, A. N., and Yano, H. (2005). "Novel high-strength biocomposites based on microfibrillated cellulose having nano-order-unit web-like network structure," *Applied Physics A* 80, 155-159.

Nishio, Y., Hirose, N., and Takahashi, T. (1989). "Thermal-analysis of cellulose poly(ethylene oxide) blends," *Polymer Journal* 21(4), 347-351.

Pothan, L. A., Oommen, Z., and Thomas, S. (2003). "Dynamic mechanical analysis of banana fiber reinforced polyester composites," *Composites Science and Technology* 63(2), 283-293.

Roohani, M., Habibi, Y., Belgacem, N. M., Ebrahim, G., Karimi, A. N., and Dufresne, A. (2008). "Cellulose whiskers reinforced polyvinyl alcohol copolymers nanocomposites," *European Polymer Journal* 44(8), 2489-2498.

Rowell, R. M., Han, J. S., and Rowell, J. S. (2000). "Characterisation and factors effecting fiber properties," *Natural Polymers and Agrofibers Based Composites*, 115-134.

Samir, M. A. S. A., Alloin, F., and Dufresne, A. (2005). "Review of recent research into cellulosic whiskers, their properties and their application in nanocomposite field," *Biomacromolecules* 6(2), 612-626.

Samir, M. A. S. A., Alloin, F., Paillet, M., and Dufresne, A. (2004). "Tangling effect in fibrillated cellulose reinforced nanocomposites," *Macromolecules* 37(11), 4313-4316.

Sawatari, C., and Kondo, T. (1999). "Interchain hydrogen bonds in blend films of poly(vinyl alcohol) and its derivatives with poly(ethylene oxide)," *Macromolecules*, 32(6), 1949-1955.

Schartel, B., Wendling, J., and Wendorff, J. H. (1996). "Cellulose poly(vinyl alcohol) blends. 1. Influence of miscibility and water content on relaxations," *Macromolecules* 29(5), 1521-1527.

Schultz, T. P., McGinnis, G. D., and Bertran, M. S. (1985). "Estimation of cellulose crystallinity using Fourier transform infrared spectroscopy and dynamic thermogravimetry," *Journal of Wood Chemistry and Technology* 5(4), 543-557.

Schwanninger, M., Rodrigues, J., Pereira, H., and Hinterstoisser, B. (2004). "Effects of short-time vibratory ball milling on the shape of FT-IR spectra of wood and cellulose," *Vibrational Spectroscopy* 36(1), 23-40.

Seydibeyoğlu, M. Ö., and Oksman, K. (2008). "Novel nanocomposites based on polyurethane and micro fibrillated cellulose," *Composites Science and Technology*, 68(3-4), 908-914.

Shibayama, M., Yamamoto, T., Xiao, C. F., Sakurai, S., Hayami, A., and Nomura, S. (1991). "Bulk and surface characterization of cellulose poly(vinyl alcohol) blends by Fourier-transform infrared-spectroscopy," *Polymer* 32(6), 1010-1016.

Siró, I., and Plackett, D. (2010). "Microfibrillated cellulose and new nanocomposite materials: A review," *Cellulose* 17(3), 459-494.

Tadoroko, H. (1959). "Infrared studies of polyvinyl alcohol by deuteration of its OH groups," *Bulletin of the Chemical Society of Japa*, 32(11), 1252-1257.

Tímár-Balázsy, A., and Eastop, D. (2002). *Chemical Principles of Textile Conservation*. Butterworth-Heinemann.

Yu, L., Dean, K., and Li, L. (2006). "Polymer blends and composites from renewable resources," *Progress in Polymer Science* 31(6), 576-602.

Zhu, B., Li, J. C., He, Y., and Inoue, Y. (2003). "Studies on binary blends of poly (3-hydroxybutyrate-co-3-hydroxyhexanoate) and natural polyphenol catechin: Specific interactions and thermal properties," *Macromolecular Bioscience* 3(5), 258-267.

Zimmermann, T., Bordeanu, N., and Strub, E. (2010). "Properties of nanofibrillated cellulose from different raw materials and its reinforcement potential," *Carbohydrate Polymers* 79(4), 1086-1093.

Zimmermann, T., Pöhler, E., and Geiger, T. (2004). "Cellulose fibrils for polymer reinforcement," *Advanced Engineering Materials* 6(9), 754-761.

Article submitted: November 27, 2010; Peer review completed: February 24, 2011; Revised version received and accepted: March 22, 2011; Published: March 24, 2011.

Publication D

Gregorova, A., Wimmer, R., Hrabalova, M., Koller, M., Ters, T. & Mundigler, N.

Effect of Surface Modification of Beech Wood Flour on Mechanical and Thermal Properties of Poly (3-hydroxybutyrate)/Wood Flour Composites.

Holzforschung, 2009, 63, 565-570.

Effect of surface modification of beech wood flour on mechanical and thermal properties of poly (3-hydroxybutyrate)/wood flour composites

Adriana Gregorova[1,*], Rupert Wimmer[2], Marta Hrabalova[3], Martin Koller[4], Thomas Ters[1] and Norbert Mundigler[3]

[1] Institute of Wood Science and Technology, University of Natural Resources and Applied Life Sciences, Vienna, Austria
[2] Faculty of Forest Sciences and Forest Ecology, Georg-August-University Göttingen, Göttingen, Germany
[3] Institute for Natural Materials Technology, University of Natural Resources and Applied Life Sciences, Tulln, Austria
[4] Institute of Biotechnology and Biochemical Engineering, Graz University of Technology, Graz, Austria

*Corresponding author.
Institute of Wood Science and Technology, University of Natural Resources and Applied Life Sciences, Vienna, Peter Jordan Strasse 82, A-1190 Vienna, Austria
E-mail: adriana.gregorova@boku.ac.at

Abstract

Poly (3-hydroxybutyrate) (PHB), a biodegradable polymer from the polyhydroxyalkanoate biopolyester class, was filled with 20% beech wood flour (WF) to form completely biodegradable films. In the present study, the influence of surface modification of wood flour was investigated on the interfacial adhesion of PHB/WF composites. In addition to a hydrothermal pretreatment, sodium hydroxide and stearic acid were used as surface modifiers. Direct measurement of interfacial adhesion was carried out by mechanical testing and dynamic mechanical analysis. Thermal properties, degree of crystallinity of PHB/WF composites were determined by differential scanning calorimetry. Effects of sodium hydroxide and stearic acid treatment on the adhesion of PHB/WF interface were feeble when no hydrothermal pretreatment was applied. Nevertheless, surface modifiers applied on hydrothermally pretreated WF significantly improved the WF/PHB interface adhesion.

Keywords: biodegradable polymer; composites; mechanical properties; thermal properties; wood flour.

Introduction

Poly (3-hydroxybutyrate) (PHB), the homopolyester of 3-hydroxybutyrate, is a biodegradable aliphatic polyester that can be produced biotechnologically by numerous naturally occurring microorganisms. PHB is the best investigated compound among all polyhydroxyalkanoates (PHAs). For living organisms, PHAs serve as intracellular storage materials for carbon and energy, which are degraded at restricted availability of external carbon, thus providing the cell an advantage for survival under starvation conditions. As regards industrial utilization, PHAs are synthesized from renewable resources and contribute to conserve fossil feedstocks (Koller et al. 2009). PHB is commercially available with some potential applications for all-day commodity items and in agriculture (Holmes 1985; Braunegg et al. 1998; Mohanty et al. 2000). Moreover, PHB is a promising semicrystalline, hydrophobic thermoplastic polymer for packaging because of its low water vapor permeability and biodegradability (Poley et al. 2005; Bucci et al. 2007). However, its applicability is limited by some drawbacks, such as high production costs, brittleness, slow crystallization, and low temperature stability in the molten state (Dacko et al. 2006). This can be overcome by the synthesis of copolymers, for instance poly-3-hydroxybutyrate-co-3-hydroxyvalerate (PHB/HV). Such PHA co- and terpolyesters feature advanced material characteristics, processability, and lower crystallinity (Koller et al. 2007). As the production of co- and terpolyesters classically requires additional expensive co-substrates (Koller et al. 2009), improvements of mechanical and thermal properties of the PHB homopolyester might be achieved by compounding with other polymers or by the incorporation of fillers, such as wood flour (WF) (Gatenholm et al. 1992; Chen et al. 2002; Bergmann and Owen 2003).

WF is a material with the following characteristics: abundantly available, renewable, non-abrasive, low dense, and relatively cost-effective. It is commercially produced mainly from industrial byproducts, such as sawdust and planer shavings. WF has attracted much attention as filler in polyolefins, but its incorporation in biodegradable polymers – such as polylactic acid (PLA), and PHB – has also been reported (Reinsch and Kelley 1997; Bhavesh et al. 2008; Sykacek et al. 2009). These studies illustrated that WF can be incorporated into the matrix of these biopolymers with an advantage of supporting biodegradability at decreased price of the resulting product. Because of the low compatibility between WF and polymer matrix, there is a necessity to improve the interfacial adhesion between the two different materials (Fernandes et al. 2004; Wu 2006).

The purpose of this study was (1) to investigate WF modifications, such as hydrothermal pretreatment, and treatment with sodium hydroxide and/or stearic acid, to improve interfacial adhesion between WF and the PHB matrix, and (2) to assess the resulting mechanical and thermal properties of PHB/WF composites. The hypothesis was that proper selection of adequate WF modification would enhance interfacial adhesion and thus lead to improved material performance. The expectation was that hydrothermal pretreatment and sodium hydroxide treatment should alter chemical composition of cell wall constituents in such a way that a better compatibility accrue to the hydrophobic polymer matrix (Sivonen et al. 2002; Bruno et al. 2008; Ishikura and Nakano 2008). The treatment with stearic acid was also applied in order to restrain agglomerating of wood particles on the surface. To the best of our knowledge, the application of stearic acid as surface promoter for WF has not yet been reported, though it is commonly used as a disperser for mineral fillers (Demjen et al. 1998; Mareri et al. 1998; Li and Weng 2008). The response of WF to the modification will be observed by FT-IR spectroscopy. Dynamic mechanical analysis (DMA) will also be applied to characterize the surface treatments on resulting PHB/WF interface adhesion. DMA is a sensitive technique for observation of viscoelastic properties of composites as a function of the temperature as well as to determine interfacial bonding of composites (Edie et al. 1993; Keusch and Haessler 1997; Huda et al. 2006).

Materials and methods

Technical WF from European beech (Fagus sylvatica L.) with a particle size of 120 μm was supplied by Lindner Mobilier s.r.o. Madunice, Slovakia. PHB was received as homopolymer powder from Graz University of Technology, Institute of Biotechnology and Biochemical Engineering, Graz, Austria.

Wood flour modification

Hydrothermal pretreatment Thermal modification took place under hot and steamy conditions at 100°C for 7 h in a laboratory-scale autoclave. After the treatment, the WF was vacuum dried at 60°C for 48 h.

Alkali treatment Beech WF was immersed in sodium hydroxide solution (5% w/v) for 2 h at room temperature. The suspension was further filtered. The residue was washed with distilled water and its pH was adjusted to 9.0 with acetic acid (Huda et al. 2008). Finally, the WF was washed with distilled water again and vacuum dried at 60°C for 48 h.

Stearic acid treatment Beech WF was immersed in a 0.07 M solution of stearic acid in toluene and stirred for 48 h at room temperature. Then WF was washed with ethanol and vacuum dried at 60°C for 48 h. Weight percentage gain (WPG) was calculated as % WPG = 100 ((W_2–W_1)/W_1), where, W_1 is the weight of the dry sample prior to the modification and W_2 is the weight after modification.

FT-IR spectroscopy

The dried untreated and treated beech WF were embedded in KBr pellets and analyzed with a Vertex 70 FT-IR spectrometer (Bruker Optik) equipped with a Miracle-Diamond ATR (Pike). Spectra were collected in the range between 4000 and 600 cm^{-1} with an accumulation of 32 scans and a resolution of 2 cm^{-1}.

Table 1 Description and nomenclature of PHB and PHB films prepared with wood flour, PHB/WF.

Sample	Compositions (% by wt)
PHB	100% PHB
PHB/WF	PHB+20% untreated WF
PHB/AT-WF	PHB+20% alkali treated WF
PHB/SA-WF	PHB+20% stearic acid treated WF
PHB/HT-WF	PHB+20% hydrothermally pretreated WF
PHB/HT-WF-AT	PHB+20% HT-WF followed by alkali treatment
PHB/HT-WF-SA	PHB+20% HT-WF followed by stearic acid treatment

Processing of composites

PHB powder and WF and modified WF were dried at 60°C under vacuum for 4 h before processing. Then, PHB powder was mixed with WF manually according to the composition given in Table 1. Films (0.18 mm thick) of the mixtures and neat PHB were prepared in laboratory hydraulic hot press (Collin) by compression molding at 175°C for 1 min without pressure and 2 min at 10 MPa pressure at the same temperature. Then, the samples were cooled down under pressure to 30°C. The PHB films were subsequently disintegrated into small pieces and the compression molding procedure was repeated in order to obtain homogeneous films. The composite samples (150 mm × 100 mm with 18 mm thickness) were cut to desired shapes for mechanical and DMA analysis and stored for 3 weeks at standard conditions (23°C, 50% RH) prior to testing.

Differential scanning calorimetry (DSC)

An approximately 10 mg sample was placed in a sealed aluminum pan and was analyzed under a constant nitrogen flow of 60 ml min^{-1} in a DSC 200 F3, Netzsch instrument. Indium was the calibration reference. The first heating cycle was between 30°C and 210°C at a scan rate of 20°C min^{-1}, followed by cooling to -50°C at 20°C min^{-1} to determine crystallization. The second heating cycle took place between -50°C and 210°C at 20°C min^{-1}. Melting temperature (T_m) and enthalpy of melting (ΔH_m) were determined from the endothermic peak and the temperature of crystallization (T_c), with the enthalpy of crystallization (ΔH_c) measured from the exothermic peak. Crystallinity (X_c) of PHB composites was defined according the following formula:

$$X_c(\%) = \frac{\Delta H_m}{\Delta H_m^0} \times 100 \qquad (1)$$

where ΔH_m is the experimentally determined melting enthalpy, and ΔH_m^0 is the melting enthalpy of the 100% crystalline polymer (146 J/g for PHB homopolymer) (Barham et al. 1984).

Mechanical testing

Tensile strength, elongation at break, and Young's modulus were determined on a Zwick Type BZ1 mechanical testing machine. The size of the rectangular testing samples was 40 × 10 × 0.18 mm^3. Grip clearance was 25 mm and crosshead speed was 2 mm min^{-1}. All mechanical parameters were derived from averaging five experimental runs for each film sample.

Scanning electron microscopy (SEM)

The instrument used was the Tesla BS 300 SEM. All samples were coated with gold prior to the examination of morphology.

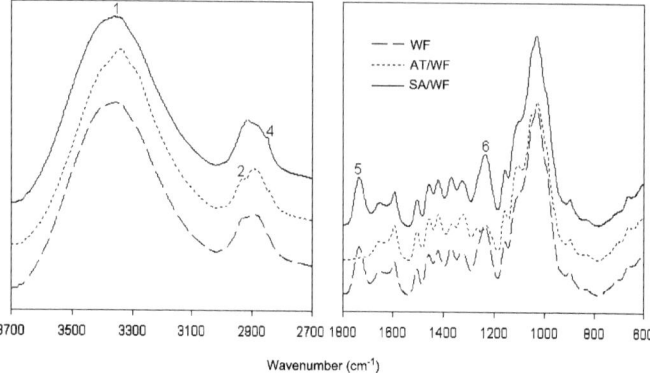

Figure 1 FT-IR-ATR spectra of untreated wood flour (WF), alkali treated WF (AT-WF), and stearic acid treated WF (SA-WF) in the spectral region from 3700 to 2700 cm⁻¹ and in the fingerprint region from 1800 to 600 cm⁻¹ (no absorbance scale is given because the spectra were shifted parallel to the wavenumber axis, numbers see text).

Dynamic mechanical properties (DMA)

The viscoelastic properties of PHB films, the storage modulus (E'), and the mechanical loss factor (tan $\delta = E''/E'$) were measured in a Netzsch DMA 242 C instrument in tension mode (Karin et al. 2006). Strips were cut from the films at the size of 10×6× 0.18 mm³. Temperature range was -20°C to +100°C, oscillation frequency was 1 Hz, and heating rate was 3°C min⁻¹.

Results and discussion

The most remarkable weight loss (WPG) was recorded after alkali treatment (-17.8%) and also in combination with hydrothermal pretreatment (-17.6%). The chemistry of alteration was confirmed by FT-IR measurement as shown in Figure 1. After alkaline treatment, the spectra of the modified beech wood flour (AT-WF) showed an increase of the broad band around 3362 cm⁻¹ (1) indicating an increase of OH-groups. The band at 1735 cm⁻¹ (5) corresponding to the C=O stretching vibrations in acetic acid esters of xylans disappeared. This indicates that the hemicelluloses are completely deacetylated. The loss of acetyl groups can also be seen at 2930 cm⁻¹ (2), which is the band of stretching vibrations of the methyl group, and further at 1235 cm⁻¹ (6), the region of the C-O stretching vibrations of lignin and xylans. The lignin bands in the fingerprint region show almost no alteration. It can be assumed that the alkaline treatment influenced only the hemicelluloses. On the other hand, untreated WF and hydrothermally pretreated WF (HT-WF) revealed low WPG (0.4% and 1.1%, respectively) after stearic acid treatment. Also, the FT-IR spectra indicated just small structural changes. In the spectra of SA-WF, small bands appeared at 2918 and 2853 cm⁻¹ originating from the stretching and asymmetric valence vibrations of the methylene groups in the carbon chain of stearic acid. The small amount of absorbed stearic acid might be attributed to the adsorption of stearic acid on the surface of WF.

Thermal properties

The crystallization and melting behavior of the neat PHB and PHB/WF composites were studied by DSC analysis by employing three thermal cycles between -50°C and 210°C. The data are summarized in Table 2. Figure 2 shows DSC thermograms recorded for neat PHB and PHB/WF samples. In general, PHB polymer degrades above its melting temperature and its average molecular weight decreases. Moreover, the presence of prodegradants – such as residual water, acetic acid, and stearic acid appearing from surface treatment of WF – might also conduce to degradation of the PHB/WF composites. Therefore, the intention was to determine the effect of higher temperature on crystallization and melting behavior of the tested samples. It is visible from Table 2 that

Table 2 Thermal data obtained by DSC measurements for PHB and PHB/WF films.

	First heating cycle			Cooling cycle		Second heating cycle				
Sample	ΔH_m (J/g)	T_m (°C)	X_c (%)	ΔH_c (J/g)	T_c (°C)	$\Delta H_c'$ (J/g)	T_c' (°C)	ΔH_m (J/g)	T_m (°C)	X_c (%)
PHB	96.9	181	66	61.3	72	0.5	48	87.3	175	60
PHB/WF	77.5	186	53	53.8	76	–	–	71.2	174	49
PHB/AT-WF	77.6	185	53	57.4	81	–	–	71.8	175	49
PHB/SA-WF	76.7	189	53	51.5	78	–	–	69.4	174	48
PHB/HT-WF	73.1	180	50	54.3	76	–	–	70.0	175	48
PHB/HT-WF-AT	78.8	181	54	54.2	73	–	–	68.7	174	47
PHB/HT-WF-SA	71.7	178	49	52.1	76	–	–	68.8	174	47

Figure 2 DSC thermograms (cooling and second heating cycle) of PHB (—) and PHB/WF (----) films; the scan rate was 20°C min⁻¹ throughout.

T_m obtained from the first heating cycle for the PHB composites with untreated WF, AT-WF, and SA-WF was higher than that of neat PHB. This tendency was not registered for HT-WF. The decline of crystallinity was noted for all PHB/WF composites. This effect was even more pronounced for the HT samples (PHB/HT-WF), particularly for SA-WF (PHB/HT-WF-SA). Enhancement of filler dispersion and interfacial adhesion is known to reduce crystallization of polymer matrix (Sanchez-Garcia et al. 2008). All the PHB/WF composites were found to crystallize at higher temperatures than neat PHB (cooling cycle). The numerical values obtained from the second heating cycle provide lower values of melting temperatures and enthalpy for all samples. This may be due to the thermal degradation of samples during DSC analysis as indicated also by the yellowish appearance of samples after thermal measurement.

With regard to high crystallinity of PHB, it is impossible to evaluate glass transition temperature by the DSC technique. Therefore, T_g was determined by DMA due to its higher sensitivity to detect molecular relaxations (Chartoff et al. 1994).

Mechanical properties

Table 3 shows mechanical properties of neat PHB and PHB/WF films. The composites containing treated WF, SA-WF in particular, induced the improvement of the tensile strength and elongation at break in comparison to

Table 3 Mechanical properties of PHB and PHB/WF films (average±SD, $n=5$).

Sample	Tensile strength (MPa)	Elongation at break (%)	Young's modulus (MPa)
PHB	36.4±2.8	2.1±0.3	2990±209
PHB/WF	30.9±2.9	1.2±0.2	3160±245
PHB/AT-WF	32.2±1.1	1.5±0.1	3240±92
PHB/SA-WF	39.6±1.7	1.7±0.2	3420±77
PHB/HT-WF	33.8±2.3	1.3±0.1	3320±184
PHB/HT-WF-AT	32.8±2.9	1.3±0.3	3310±25
PHB/HT-WF-SA	36.5±2.9	1.4±0.1	3450±258

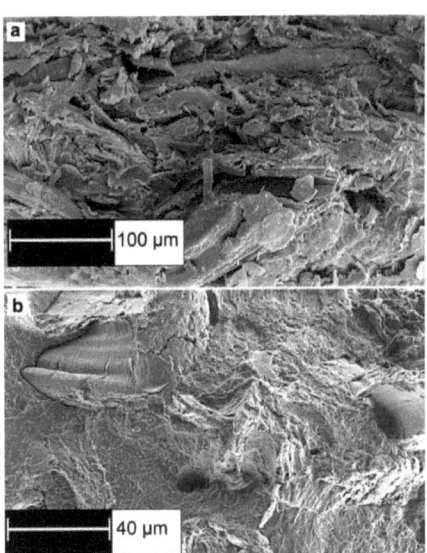

Figure 3 SEM micrographs of microstructures of (a) PHB/WF (PHB with untreated wood flour) and (b) PHB/HT-WF-SA (PHB with hydrothermally pretreated wood flour treated by stearic acid) films.

PHB composite with untreated WF. Young's moduli of the PHB composites with treated WF were clearly higher than the neat PHB samples and the PHB composites filled with untreated WF, respectively. The composites with SA-WF exhibited the highest Young's moduli and tensile strength. This may reflect the reinforcing effect of the treatment through improved interfacial adhesion between the PHB matrix and WF.

Figure 3 illustrates the fracture surface of the PHB/WF and PHB/HT-WF-SA composites, respectively. Figure 3a shows that the interfacial binding between untreated WF and PHB matrix was very poor. Compared to Figure 3b, the interface between HT-WF treated by SA and the PHB matrix was improved greatly.

Dynamic mechanical analysis (DMA)

The temperature-dependent curves of storage modulus and loss factor are presented in Figure 4. Storage modulus characterizes the ability of the polymer to store energy and reflects the stiffness of the measured sample. As provided, the incorporation of both untreated and treated WF have increased storage modulus of PHB composites and this effect was most pronounced for the HT-WF. This finding may be attributed to a reinforcing effect of the WF, although the even higher increase in stiffness for HT-WF suggests improved contribution of the filler to PHB matrix interaction (Singha and Mohanty 2007). As provided in Figure 4a, the PHB/HT-WF-SA composite had the highest storage modulus compared to other PHB/WF composites. This suggests that a combined hydrothermal-stearic acid treatment may be the most promising

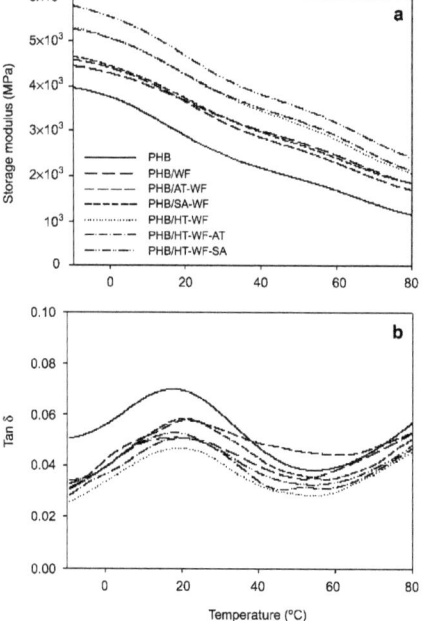

Figure 4 Temperature dependence of (a) storage modulus, (b) loss factor of PHB and its composites.

modification for beech WF to obtain PHB/WF composites with high stiffness. As shown in Figure 4a and Table 4, the storage modulus for all PHB composites decreased as temperature increased. However, PHB/WF composites displayed higher stiffness over the entire temperature span in comparison to neat PHB, especially the PHB films filled with HT-WF. At higher temperatures the reinforcing effect of WF is even more pronounced because of its ability to restrict motions of the PHB chains. At 80°C, the PHB/HT-WF-SA composite showed a 111% increase of the storage modulus compared to neat PHB (Table 4). The other PHB/WF composite types showed improvements between 48% and 85% for the storage modulus at 80°C relative to neat PHB. The significant increase in the storage modulus of PHB composites with 20 wt% HT-WF modified with stearic acid may be attributed to increased interfacial adhesion between the WF and the PHB matrix caused by the used treatments. These results are consistent with the mechanical behavior of PHB films.

Glass transition temperature (T_g) was determined as the peak temperature of the loss factor (tan δ) curve (Table 4) corresponding to the transition midpoint (Menard 1999). Incorporation of 20% WF – untreated and treated – slightly shifted T_g to higher temperatures. This may be due to the recession of the polymer molecular chains mobility induced by hindering of WF reinforcements and leading to the reduction and shift of height of tan δ peak. As provided in Figure 4b, WF incorporation slightly lowered the intensity of the tan δ peak of all the PHB/WF composite types compared to the neat PHB. This effect was most pronounced with the HT-WF. Fay et al. (1991) claimed that there was a consistent proportion between the reduction in loss factor and a reduction in friction of intermolecular polymer chains. The findings suggest that the HT-WF decreased damping in transition region thus reflecting imperfection in the elasticity and reduction of the internal friction.

Conclusions

PHB films reinforced with 20% WF were prepared by melt pressing. The incorporation of all WF increased crystallization temperature. Following the mechanical testing it can be concluded that the incorporation of WF increased Young's modulus. The stearic acid treatment retained tensile strength compared to neat PHB film. The highest value of the storage modulus at 20°C was determined for the PHB composite containing HT-WF, modified with stearic acid (HT-WF-SA, 4.68 GPa). This provided a significant improvement compared to the PHB composites based on untreated WF (3.67 GPa). The observed increase in storage modulus of PHB composites filled with HT-WF indicated a better interfacial bonding between filler and the PHB matrix. Increase in glass transition temperature of PHB/WF due to the treatments reflects the lower segmental motion of polymer molecules, the reinforcement effect, and the improved interfacial adhesion. Based on the obtained results, it is concluded that the most effective treatment for beech WF was the hydrothermal pretreatment followed by a treatment with stearic acid. Hydrothermal pretreatment increased the subsequent treatability of WF, and stearic acid treatment ensured better dispersion of WF in the PHB matrix. The use of PHB/WF composites is attractive mainly due to the biodegradability and renewability of their constituents.

Acknowledgements

This work was supported by the Austrian Science Fund FWF (project no. L319-B16).

References

Barham, P.J., Keller, A., Otun, E.L., Holmes, P.A. (1984) Crystallization and morphology of a bacterial thermoplastic: poly-3-hydroxybutyrate. J. Mater. Sci. 19:2781–2794.

Table 4 Dynamic mechanical properties of PHB and PHB/WF films.

Sample	E' (MPa)				T_g (°C)	Tan δ at peak
	0°C	20°C	40°C	80°C		
PHB	3751	2883	2179	1143	17.7	0.070
PHB/WF	4286	3667	3007	1839	18.3	0.051
PHB/AT-WF	4403	3650	2854	1693	20.7	0.058
PHB/SA-WF	4446	3725	2975	1844	21.0	0.058
PHB/HT-WF	5063	4256	3447	2072	18.9	0.047
PHB/HT-WF-AT	5059	4264	3498	2125	18.1	0.053
PHB/HT-WF-SA	5534	4677	3815	2413	19.7	0.051

Bergmann, A., Owen, A. (2003) Hydroxyapatite as a filler for biosynthetic PHB homopolymer and P(HB-HV) copolymers. Polym. Int. 52:1145–1152.

Bhavesh, L.S., Selke, S.E., Walters, M.B., Heiden, P.A. (2008) Effects of wood flour and chitosan on mechanical, chemical and thermal properties of polylactide. Polym. Compos. 29:655–663.

Braunegg, G., Lefebvre, G., Genser, K.F. (1998) Polyhydroxyalkanoates, biopolyesters from renewable resources: physiological and engineering aspects. J. Biotechnol. 65:127–161.

Bruno, E., Graca, J., Pereira, H. (2008) Extractive composition and summative chemical analysis of thermally treated eucalypt wood. Holzforschung 62:344–351.

Bucci, D.Z., Tavares, L.B.B., Sell, I. (2007) Biodegradation and physical evaluation of PHB packaging. Polym. Test. 26:908–915.

Chartoff, R.P., Weissman, P.T., Sircar A. (1994) The application of dynamic mechanical methods to Tg determination in polymers. In: Assignment of the Glass Transition. Ed. Seyler, R.J. ASTM International, Atlanta, GA, USA. pp. 88–107.

Chen, Ch., Fei, B., Peng, S., Zhuang, Y., Dong, L., Feng, Z. (2002) Nonisothermal crystallization and melting behavior of poly(3-hydroxybutyrate) and maleated poly(3-hydroxybutyrate). Eur. Polym. J. 38:1663–1670.

Dacko, P., Kowalczuk, M., Janeczek, H., Sobota, M. (2006) Physical properties of the biodegradable polymer compositions containing natural polyesters and their synthetic analogues. Macromol. Symp. 239:209–216.

Demjen, Z., Pukanszky, B., Nagy, J. (1998) Evaluation of interfacial interaction in polypropylene/surface treated $CaCO_3$ composites. Compos. Part A 29A:323–329.

Edie, D.D., Kennedy, J.M., Cano, R.J., Ross, R.A. (1993) Evaluating surface treatment effects on interfacial bond strength using dynamic mechanical analysis. In: Composite Materials. Fatigue and Fracture. Eds. Stinchcomb, W.W., Ashbaugh, N.E. ASTM International, Indianapolis, IN, USA. pp. 419–430.

Fay, J.J., Murphy, C.J., Thomas, D.A., Sperling, L.H. (1991) Effect of morphology, cross-link density, and miscibility on inter-penetrating polymer network damping effectiveness. Polym. Eng. Sci. 31:1731–1741.

Fernandes, E.G., Pietrini, M., Chiellini, E. (2004) Bio-based polymeric composites comprising wood flour as filler. Biomacromolecules 5:1200–1205.

Gatenholm, P., Kubat, J., Mathiasson, A. (1992) Biodegradable natural composites. I. Processing and properties. J. Appl. Polym. Sci. 45:1667–1677.

Holmes, P.A. (1985) Applications of PHB – a microbially produced biodegradable thermoplastic. Phys. Technol. 16:32–36.

Huda, M.S, Drzal, L.T., Mohanty, A.K., Misra, M. (2006) Chopped glass and recycled newspaper as reinforcement fibers in injection molded poly(lactic acid) (PLA) composites: a comparative study. Compos. Sci. Technol. 66:1813–1824.

Huda, M.S., Drzal, L.T., Mohanty, A.K., Misra, M. (2008) Effect of fiber surface-treatments on the properties of laminated biocomposites from poly(lactic acid) (PLA) and kenaf fibers. Compos. Sci. Technol. 68:424–432.

Ishikura, Y., Nakano, T. (2008) Compressive stress-strain properties of natural materials treated with aqueous NaOH. Holzforschung 62:448–452.

Karin, M., Bogren, E., Gamstedt, K., Neagu, R.C., Kerholm, M.A., Lindström, M. (2006) Dynamic-mechanical properties of wood-fiber reinforced polylactide: experimental characterization and micromechanical modeling. J. Thermoplast. Compos. 19:613–637.

Keusch, S., Haessler, R. (1997) Influence of surface treatment of glass fibres on the dynamic mechanical properties of epoxy resin composites. Compos. Part A 30:997–1002.

Koller, M., Hesse, P.J., Bona, R., Kutschera, C., Atlic, A., Braunegg, G. (2007) Biosynthesis of high quality polyhydroxyalkanoate co- and terpolyesters for potential medical application by the Archaeon haloferax mediterranei. Macromol. Symp. 253:33–39.

Koller, M., Hesse, P.J., Bona, R., Kutschera, C., Atlic, A., Braunegg, G. (2007) Current advances in cost efficient polyhydroxyalkanoate production. Curr. Trends Biotechnol. 3:1–13.

Li, Y., Weng, W. (2008) Surface modification of hydroxyapatite by stearic acid: characterization and in vitro behaviors. J. Mater. Sci: Mater. Med. 19:19–25.

Mareri, P., Bastide, S., Binda, N., Crespy, A. (1998) Mechanical behaviour of polypropylene composites containing mineral filler: effect of filler surface treatment. Compos. Sci. Technol. 58:747–752.

Menard, K.P. Dynamic Mechanical Analysis. A Practical Introduction. CRC Press LLC, Boca Raton, FL, USA, 1999.

Mohanty, A.K., Misra, M., Hinrichsen, G. (2000) Biofibres, biodegradable polymers and biocomposites: an overview. Macromol. Mater. Eng. 276/277:1–24.

Poley, L.H., Silva, M.G., Vargas, H., Siqueira, M.O., Sanchez, R. (2005) Water and vapor permeability at different temperatures of poly (3-hydroxybutyrate) dense membranes. Polimeros 15:22–26.

Reinsch, V.E., Kelley, S. (1997) Crystallization of poly (hydroxybutyrate-co-hydroxyvalerate) in wood fiber-reinforced composites. J. Appl. Polym. Sci. 64:1785–1796.

Sanchez-Garcia, M.D., Gimenez, E., Lagaron, J.M. (2008) Morphology and barrier properties of solvent cast composites of thermoplastic biopolymers and purified cellulose fibers. Carb. Polym. 71:235–244.

Singha, S., Mohanty, A.K. (2007) Wood fiber reinforced bacterial bioplastic composites: fabrication and performance evaluation. Compos. Sci. Technol. 67:1753–1763.

Sivonen, H., Maunu, S.L., Sundholm, F., Jämsä, S., Viitaniemi, P. (2002) Magnetic resonance studies of thermally modified wood. Holzforschung 56:648–654.

Sykacek, E., Schlager, W., Mundigler, N. (2009) Compatibility of softwood flour and commercial biopolymers in injection molding. Polym. Compos. in press, 7 April 2009 published online http://www3.interscience.wiley.com.

Wu, Ch.-S. (2006) Assessing biodegradability and mechanical, thermal, and morphological properties of an acrylic acid-modified poly(3-hydroxybutyric acid)/wood flours biocomposite. J. Appl. Polym. Sci. 102:3565–3574.

Received December 11, 2008. Accepted April 16, 2009.
Previously published online June 29, 2009.

9 CV

Personal data

Name	Marta Hrabalova
Date of Birth	15. 10. 1981
Place of Birth	Prilepy
Country	Czech republic
Nationality	Czech
Marital status	Unmarried
University	BOKU

Education

2007-today **Universität für Bodenkultur Wien, Department of Material Sciences and Process Engineering, IFA Tulln;** PhD studentship on viscoelastic, thermal and optical properties of biodegradable composites.

2009 **STMS-COST ACTION E50 CEMARE** (Short-term scientific mission on the base of approved project; EMPA, Zurich, Switzerland); Optimizing nano-fibrillated cellulose composites with biopolymer matrices.

2002-2007 **Tomas Bata University in Zlin, Faculty of Technology**; Polymer engineering;

2005 Bachelor thesis themed: Photodegradation of β-nucleated polypropylene; Bachelor degree of polymer engineering.

2007 **Blaise Pascal University of Clermont-Ferrand**; France; Educational stay in scope of Erasmus/Socrates (3 months).

2007 Diploma thesis themed: Photodegradation of β-nucleated polypropylene: effect of structure; Master degree of polymer engineering.

2001-2002 **AICL** (Australian Institut of Commerce and Language in Sydney);

1997-2001 **Gymnazium Zlin Lesni Ctvrt** (School leaving exam: literature, English Language, mathematics, computer graphics).

Vienna

21. 3. 2011

List of Publications:

Hrabalova, M., Schwanninger, M., Gregorova, A., Wimmer, R., Zimmermann, T., Mundigler, N., (2011); Fibrillation of Flax and Wheat Straw Cellulose and Its Effect on Thermal, Morphological and Viscoelastic Properties of Poly(vinylalcohol)/Fibre Composites; *BioResources*, 6, 2, 1631-1647.

Gregorova, A., Hrabalova, M., Kovalcik, R., Wimmer, R., (2011); Surface Modification of Spruce Wood Flour and Effects on the Dynamic Fragility of PLA/Wood Composites; *Polymer Engineering and Science*, 51, 1, 143-150

Hrabalova, M., Gregorova, A., Wimmer, R., Sedlarik, V., Machovsky, M., Mundigler, N. (2010); Effect of wood flour loading and thermal annealing on viscoelastic properties of poly(lactic acid) composite films; *Journal of Applied Polymer Science*, 118, 3, 1534–1540.

D'Amico, S., Hrabalova, M., Müller, U., Berghofer, E. (2010); Bonding of spruce wood with wheat flour glue—Effect of press temperature on the adhesive bond strength; *Industrial Crops and Products*, 31, 2, 255-260.

Sykacek, E., Hrabalova, M., Frech, H., Mundigler, N. (2009); Extrusion of five biopolymers reinforced with increasing wood flour concentration on a production machine, injection moulding and mechanical performance; *Composites Part A: Applied Science and Manufacturing*, 40, 8, 1272-1282.

Gregorova, A., Hrabalova, M., Wimmer, R., Saake, B., Altaner, C. (2009); Poly(lactide acid) Composites Reinforced with Fibers Obtained from Different Tissue Types of Picea sitchensis; *Journal of AppliedPolymer Science*, 114, 2616–2623.

Gregorova, A., Wimmer, R., Hrabalova, M., Koller, M., Ters, T., Mundigler, N. (2009); Effect of Surface Modification of Beech Wood Flour on Mechanical and Thermal Properties of Poly (3-hydroxybutyrate)/Wood Flour Composites; *Holzforschung*, 63, 565-570.

Conferences:

Hrabalova, M., Wimmer, R., Gregorova, A.; Effect of wood-flour Loading and Thermal Annealing on Dynamic-Mechanical Properties of Poly(lactic acid) Composite Films; *NARO-TECH 2009 7. Internationales Symposium „Werkstoffe aus Nachwachsenden Rohstoffen" Erfurt*, 9.-10.9. 2009, Erfurt, Germany; poster, manuscript.

Navratilova, J., Cermak, R., Hrabalova, M., Commereuc, S., Verney, V.; The Role of Molecular Weight in Photodegradation of β-nucleated polypropylene; EPF 09 European polymer congress; 12-17.7.2009, Graz, Austria; poster.

Hrabalova, M., Wimmer, R., Gregorova, A., Mundigler, N.; Light-transmitting Wood Through Resin Impregnation; EPF 09 European polymer congress; 12-17.7.2009, Graz, Austria; poster.

Gregorova, A., Hrabalova, M., Wimmer, R., Sedlarik, V.; Using maleic anhydride grafted poly(lactic acid) (PLA) reinforced with modified spruce wood flour (WF) for improved biobased composites; EPF 09 European polymer congress; 12-17.7.2009, Graz, Austria; poster.

Vychopnova, J., Hrabalova, M., Cermak, R., Mosnovska, R., Commereuc, S., Verney, V.; Photodegradation of β-nucleated polypropylene: The effect of structure parameters; Proceedings of the Polymer Processing Society 24th Annual Meeting, PPS-24, 15.-19.7. 2008, Salerno, Italy; poster.

Hrabalova, M., Wimmer, R., Gregorova, A., Mundigler, N.; Effect of Surface treatment and Preparation Method on the properties of Poly(lactic acid)-Wood Fibre Composites; Proceedings of the Polymer Processing Society 24th Annual Meeting, PPS-24, 15.-19.7. 2008, Salerno, Italy; poster.

Gregorova,A., Wimmer, R., Hrabalova, M., Koller, M., Braunegg,G.; Dynamic-mechanical and Thermal Properties of Poly(3-hydroxybutyrate)/wood flour films; Proceedings of the Polymer Processing Society 24th Annual Meeting, PPS-24, 15.-19.7. 2008, Salerno, Italy; poster.

Gregorova, A., Wimmer, R., Hrabalova, M., Mundigler, N.; Dynamic - Mechanical and Thermal Properties of Biodegradable Composites from Polylactic Acid (PLA) Reinforced with Wood Fibers; International Conference (Bio)Degradable Polymers from Renewable Resources; 18.-21.11. 2007, Vienna, Austria; poster.

Dissertations and dissertation print of the University of Natural Resources and Applied Life Sciences Vienna

Module 1: Dissertations

Imprint of this document:
© 2004 Guthmann-Peterson Publisher
Elßlergasse 17, A-1130 Vienna
Phone. +43 (0)1 877 04 26, Fax: +43 (0)1 876 40 04
Dr.-Simoneit-Straße 36, D-45473 Mülheim a. d. Ruhr
E-Mail: verlag@guthmann-peterson.de
http://www.guthmann-peterson.de

The present document template and your own documents are copyrighted texts that are only intended for writing dissertations. In your own interest and also on behalf of the university, further usage of the template, changes of the document or any other type of commercial use are not permitted.

All hard- and software names in this document are registered trademarks. This document and the sample files "template-dissertation-boku.doc", and the corresponding PDF v

Die VDM Verlagsservicegesellschaft sucht für wissenschaftliche Verlage abgeschlossene und herausragende

Dissertationen, Habilitationen, Diplomarbeiten, Master Theses, Magisterarbeiten usw.

für die kostenlose Publikation als Fachbuch.

Sie verfügen über eine Arbeit, die hohen inhaltlichen und formalen Ansprüchen genügt, und haben Interesse an einer honorarvergüteten Publikation?

Dann senden Sie bitte erste Informationen über sich und Ihre Arbeit per Email an *info@vdm-vsg.de*.

Sie erhalten kurzfristig unser Feedback!

VDM Verlagsservicegesellschaft mbH
Dudweiler Landstr. 99
D - 66123 Saarbrücken
www.vdm-vsg.de

Telefon +49 681 3720 174
Fax   +49 681 3720 1749

Die VDM Verlagsservicegesellschaft mbH vertritt

Printed by Books on Demand GmbH, Norderstedt / Germany